HOW TO
FORECAST
WEATHER

No. 1568
$16.95

HOW TO
FORECAST
WEATHER

BY DAN RAMSEY

TAB BOOKS Inc.
BLUE RIDGE SUMMIT, PA. 17214

This book is dedicated to Heather Kay who makes the sun shine on a cloudy day.

FIRST EDITION

SECOND PRINTING

Printed in the United States of America

Library of Congress Cataloging in Publication Data

Ramsey, Dan, 1945-
How to forecast weather.

Includes index.
1. Weather forecasting—Popular works.   I. Title.
QC995.4.R35   1983   551.6'3   83-8249
ISBN 0-8306-0268-2
ISBN 0-8306-0168-6 (pbk.)

Cover illustration by Al Cozzi.

Contents

Acknowledgments

I wish to express my indebtedness to many government bodies, private corporations, and associations who supplied a downpour of information, illustrations, and publications that made the complex science of meteorology into the simple hobby of weather-watching. They include:

U.S. Department of Commerce
National Oceanic and Atmospheric Administration
National Weather Service
Environmental Data Service

U.S. Department of Agriculture, Forest Service
Federal Aviation Administration
Department of the Army
Naval Education and Training Support Command

Thanks also to the weather equipment suppliers listed in the Appendix, especially Heath Company, Benton Harbor, MI.

Individually, thanks go to Donald E. Witten and Lucinda Thorpe with NOAA's National Weather Service and Rob Boatman of Heath Company.

Introduction

Benjamin Franklin aptly remarked "Some people are weather-wise, but most are otherwise."

Poor Richard was right, yet no one will deny the importance of the weather. Weather can produce economic prosperity or disaster. Weather affects business, government, and our lives whether we like them or not. But we can eliminate the adverse effects—and even profit from them—with a simple knowledge of what weather is, what it does, and how to forecast and use it.

This is a simple and entertaining book on how you can be an amateur meteorologist for fun and profit—even if you can't pronounce meteorologist. The first three chapters explain what weather is, where it comes from, and what makes summer hot and winter cold, fog lift, and rain fall. Then you'll learn about the "air masses" and "fronts" that the TV weatherman always points to on his map. And you'll consider each of the individual elements that are collectively called "the weather:" wind, barometric pressure, temperature, humidity, precipitation (rain, snow, etc.), and clouds.

Once you understand the basics of weather you'll learn to observe the weather and how it moves in and out of your life each day. (Weather folklore and fairy tales will be considered in Chapter 4—some of them useful and some bunk.) Observation gets more scientific as you learn in Chapter 5 how to make and choose your own weather shack instruments for measuring the elements. And you'll see in Chapter 6 how the National Weather Service uses state-of-the-art electronics and satellites to see over the horizon.

You will then learn in Chapter 7 how to collect and record the type of weather data needed to make accurate forecasts of weather one, two, five, and even 90 days in advance. In Chapter 8 the data is interpreted using methods as simple or as complex as you wish—from a barometer and a chart to a numerical weather computer. You'll even find out how to let the Weather Service do all the hard work while you improve on their forecasts and make them more accurate for your location.

Chapter 9 details how weather is used every day by all kinds of people—farmers, gardeners, aviators, travelers, ships' captains, pleasure boaters, business operators, recreation directors, salespeople, students, housewives, and workers.

You'll learn what comfortable weather really is and you'll find out how to put the weather to work for you.

To help you understand the many new terms you'll be reading, a comprehensive glossary of weather words is included. An extensive listing of dealers and manufacturers of weather instruments across the nation, by location is also included.

Most important, more than 230 charts, diagrams, photographs, and drawings are included to make the weather easy for you to understand and predict.

Chapter 1

The Atmosphere

Weather is the condition of the atmosphere around us measured in terms of heat, pressure, wind, and moisture. It's a warm, sunny day with light winds. It's a blizzard or a thunderstorm. It's cloudy skies and rain.

Meteorology is the science that considers—and tries to do something about—the atmosphere and its elements. It is a physical science that looks at the past and present conditions of the sky to predict or "forecast" the weather of the future.

Meteorology can be as simple or as complex as man's understanding. Intricate satellites and massive computers are used to reaffirm what the mariner already knows: "Red sky in morning, sailors take warning. Red sky at night, sailors delight." With complexity comes accuracy (usually) as man's ability to foresee weather conditions is extended. A century ago man knew little more about the what, why, and when of weather than he did a millenium before. The past few decades have multiplied his foresight many times so that he can now accurately predict weather days and even weeks in advance.

And man has learned to profit from his predictions.

MAN VERSUS THE ELEMENTS

Not that long ago, man was nearly always the victim, never the victor, of the elements (Fig. 1-1). When it rained, he got wet. When the sun shone, he was tanned. The weather decided much of what and when he would eat. Time was reckoned by "rainy seasons" and "droughts." He worshipped the elements as he worshipped life.

Then he learned that the branches of trees offered respite from the direct rays of the sun and caves kept the rain off his head. Trees and caves were difficult to move so he had a bright idea: movable shelters of wood and rock.

Observation soon taught him that watching the horizon would give him advance warning of rain and winds. He also discovered that animals acted differently just before a storm. He could then foretell or *fore cast* the weather and take advantage of the elements to improve his life. The weather became his tool.

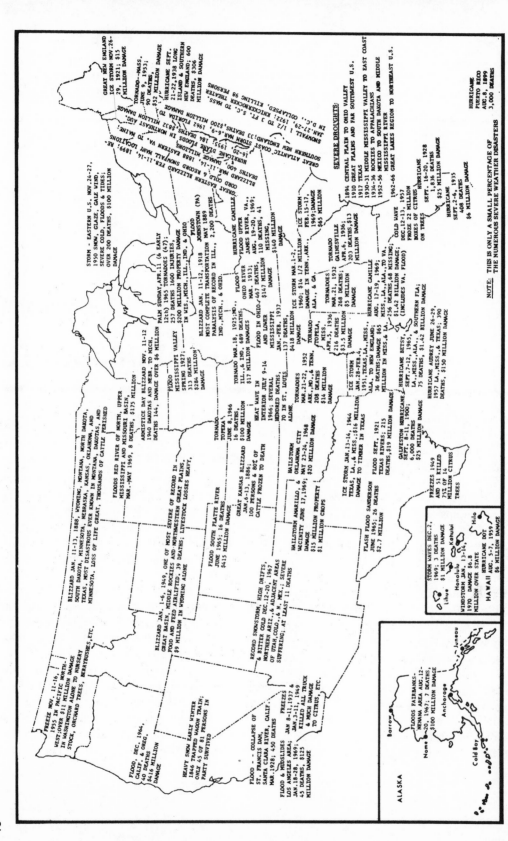

Fig. 1-1. Major weather disasters based on period of record to January, 1970.

2

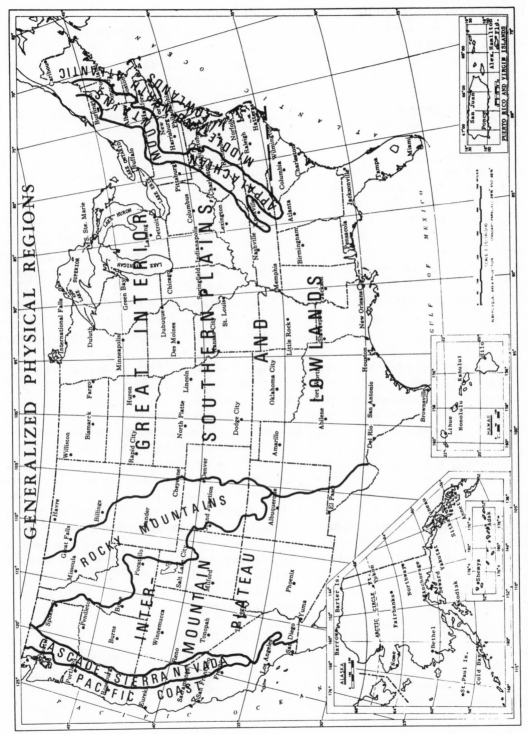

Fig. 1-2. Generalized physical regions in the U.S.

3

But his knowledge of current and future weather conditions was limited to line-of-sight. He was still a victim of any weather conditions lurking beyond the horizon. What he needed was a fleet-footed friend who lived over the hill and was willing to run over a few times a day to tell him what he saw. Friendship being what it is, man had to wait for someone to invent the telegraph before he could extend his weather observations beyond a few miles and a few hours.

Since that time, man has developed the art of weather forecasting to a science—a science that, in all levels of society, uses the elements as tools. The farmer can use the weather to decide if he should grow melons or mint. Mothers can appropriately arm their offspring with swimming trunks or snow suits. Businessmen can plan sidewalk sales. Families can plan picnics. Weather surprises still occur, but with less frequency than the days when man knew nothing of the weather.

THE CLIMATE

Climate is weather plus time. That is, the climate of an area is the type and amount of weather elements that occur over a long period of time. It may be sunny today in Seattle, Washington—but its *climate* is rainy. Climate is the history of weather in a particular location and includes averages, extremes, and trends in temperature, precipitation, humidity, wind, cloudiness, and snow cover (Fig. 1-2).

Climatology is the scientific study of climate. A *climatologist* may study the causes of climate or he may be interested in the uses of climatic data for agriculture, industry, aviation, or any one of many fields of human endeavor.

Weather—thus climate—is caused by the interaction of the sun and the Earth. If for some reason the sun were to brighten or diminish, or the Earth were to tilt or move from its place in space, our weather would react drastically and life as we know it might soon come to a screeching halt.

In order to understand weather, let's take a simplified look at the sun, the Earth, its atmosphere, and how they interact to give us warm days and growing seasons.

THE SUN

The sun is the Earth's only source of heat energy and the cause of all weather and atmospheric changes on Earth. With a surface temperature of about 11,000° Fahrenheit, the sun radiates electromagnetic energy in all directions. The Earth intercepts about one two-billionths of this energy. Most of the electromagnetic energy radiated by the sun is in the form of light waves. Only a tiny fraction is in the form of heat waves. Even so, better than 99.9 percent of the Earth's heat is derived from the sun.

The sun is a globe of gas heated to incandescence by thermonuclear reactions from within the central core (Fig. 1-3). The tremendous heat energy generated within the sun's core is transported by the radiative transfer of *photons*, which bounce from atom to atom, much like bouncing balls, through the radiative zone. Within the *convective zone*, which extends very nearly to the sun's surface, the heated gases are raised buoyantly upwards with some cooling and subsequent convective action occurring until the gases are cooled to approximately 600°K (Kelvin or Absolute) at the sun's surface.

The main body of the sun, although composed of gases, is opaque and has a well-defined, visible surface referred to as the *photosphere*. This is what we see when we look at the sun. From the photosphere all the light and heat of the sun are radiated. Above the photosphere is a more transparent gaseous layer referred to as the *chromosphere* with a thickness of about 6000 miles. Above the chromosphere is the *corona*, which may extend outward a distance of several solar diameters. As you can see in Fig. 1-3, the photosphere, chromosphere, and corona comprise what may be referred to as the *solar atmosphere*.

SOLAR ACTIVITY

Within the solar atmosphere certain more transient phenomena—referred to as *solar activity*—occur that are similar to that within the Earth's atmosphere. Each has some effect on our weather.

Solar prominences are perhaps the most beau-

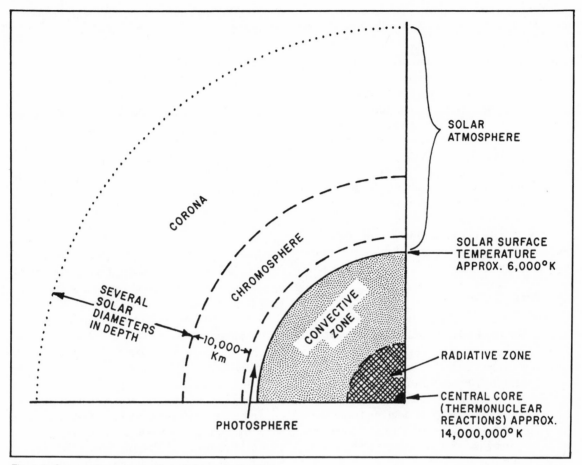

Fig. 1-3. One-quarter cross section of the sun's structure.

tifully colored appendages of the sun. They appear as great clouds of gas, sometimes resting on the sun's surface, at other times floating free with no visible connection. When viewed against the solar disc they appear as long dark filaments. They display a variety of shapes, sizes, and activity that defy general description. The more active types appear hotter than the surrounding atmosphere with temperatures near 17,000° Fahrenheit.

Sunspots appear as relatively dark areas on the surface of the sun. They may appear singly or in more complicated groups dominated by large spots near the center. Sunspots begin as small dark areas known as *pores*. These pores develop into full-fledged spots in a few days with maximum development occurring in about one to two weeks.

Sunspot *decay* consists of the spot shrinking in size. This life cycle may run from a few days for small spots to nearly 100 days for larger groups. The larger spots normally measure about 75,000 miles across. Sunspots appear to have cyclic variations in intensity, varying through a period of about eight to 17 years.

Plages are large, irregular, bright patches that surround sunspot groups. They normally appear in conjunction with solar prominences or filaments and may be systematically arranged in radial or spiral patterns. Plages are features of the lower chromosphere and are often completely or partially obscured by an underlying sunspot.

Solar flares are perhaps the most spectacular of the eruptive features associated with solar activity.

5

They appear as flecks of light that suddenly appear near activity centers, appearing as instantaneously as though a switch were thrown. They rise sharply to peak brightness in a few minutes, then decline gradually. The number of flares may increase rapidly over an area of activity. Small flare-like brightenings are always in progress during the more active phase of activity centers. The smaller flares may be classified as *subflares*. In some instances, flares may take the form of prominences, violently ejecting material into the solar atmosphere and breaking into smaller high-speed blobs or clots. Flare activity varies widely between solar activity centers and appears to be a function of the complexity of the magnetic field. The greatest flare productivity seems to be during the week or ten days when sunspot activity is at its maximum.

THE EARTH

Of the nine planets of our solar system, the Earth is the third from the sun. Its maximum distance from the sun is 94 million miles in summer; its minimum distance from the sun is 91 million miles in winter (in the Northern Hemisphere). It has an atmosphere more than 60 miles thick.

The Earth is subject to four motions in its movement through space. Only two of these motions are of any importance to meteorology: *rotational motion* (turning of the Earth on its axis) and *revolutional motion* (movement of the Earth in orbit around the sun).

In the first motion the Earth rotates on its axis once in 24 hours; one half of the Earth's surface is therefore facing the sun at all times. The side facing the sun is experiencing daylight and the side facing away from the sun is experiencing darkness, accounting for our day and night. Rotation about its axis takes place in an eastward direction. Thus, the sun rises in the east and sets in the west.

The second motion of the Earth is its revolution around the sun. This revolution and the tilt of the Earth on its axis are responsible for our seasons. The Earth makes one complete revolution around the sun in approximately 365¼ days. The Earth's axis is at an angle of 23½° to its plane of rotation. The Earth's axis points in a nearly fixed direction in space toward the North Star (Polaris) at all times.

SOLSTICES AND EQUINOXES

When the Earth is in its summer solstice, as shown in Fig. 1-4, the Northern Hemisphere is inclined at 23½° *toward* the sun. This inclination results in more of the sun's rays reaching the Northern Hemisphere than the Southern Hemisphere. On or about June 21, the sun shines *over* the North Pole down the other side to latitude 66½° (the Arctic Circle) and the most perpendicular rays of the sun are received at 23½° North latitude (the Tropic of Cancer). The Southern Hemisphere is tilted *away* from the sun at this time and the sun's rays reach only to 66½° South latitude (the Antarctic Circle) and do not go beyond this latitude. The area between the Antarctic Circle and the South Pole is in darkness; the area between the Arctic Circle and the North Pole is receiving the sun's rays for 24 hours each day.

At the equinoxes in March and September, the tilt of the Earth's axis is neither toward nor away from the sun. For this reason, the Earth receives equal numbers of the sun's rays in both the Northern and Southern Hemispheres, and the sun's rays shine most perpendicularly at the Equator.

In December, the situation is exactly reversed from that of June. The Southern Hemisphere now receives more of the sun's rays. The most perpendicular rays of the sun are received at 23½° latitude (the Tropic of Capricorn). The south polar area is completely in sunshine and the north polar area is completely in darkness.

Since the revolution of the Earth around the sun is a gradual process, the changes in the area receiving the sun's rays and the changes in seasons are gradual. However, it is customary and convenient to mark these changes by specific dates and to identify them by specific names:

March 21/vernal equinox is when the Earth's axis is perpendicular to the sun's rays. Spring begins in the Northern Hemisphere and fall begins in the Southern Hemisphere.

June 21/summer solstice is when the Earth's axis is inclined 23½ degrees toward the sun and the

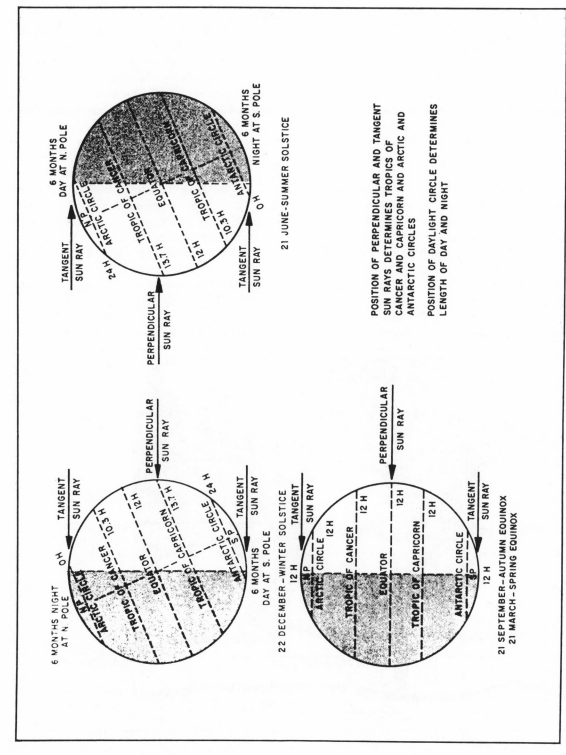

Fig. 1-4. The Earth's tilt and latitude make the difference in the amount of solar insolation received.

sun has reached its northernmost zenith at the Tropic of Cancer. Summer officially commences in the Northern Hemisphere and winter begins in the Southern Hemisphere.

September 22/autumnal equinox is when the Earth's axis is again perpendicular to the sun's rays. This date marks the beginning of fall in the Northern Hemisphere and spring in the South Hemisphere. It is also the date, along with March 21, when the sun reaches its highest position (zenith) directly over the equator.

December 22/winter solstice is when the sun has reached its southernmost zenith positions at the Tropic of Capricorn. It marks the beginning of winter in the Northern Hemisphere and the beginning of summer in the Southern Hemisphere.

In some years, the actual dates of the solstices and equinoxes vary by a day from the dates given here because the period of revolution is 365¼ days, and the calendar year is 365 days, except for leap year when it is 366 days.

ZONES

Because of its 23½° tilt and its revolution around the sun, the Earth is marked by five natural light (or heat) zones according to the zone's relative position to the sun's rays. Since the sun is *always* at its zenith between the Tropic of Cancer and the Tropic of Capricorn, this is the hottest zone. It is called either the *Equatorial Zone*, the *Torrid Zone*, the *Tropical Zone* or simply the *Tropics* (Fig. 1-5).

The zones between the Tropic of Cancer and the Arctic Circle, and between the Tropic of Capricorn and the Antarctic Circle, are the *Temperate Zones*. These zones receive sunshine all year, but less of it in their respective winters and more in their respective summers.

The zones between the Arctic Circle and the North Pole and between the Antarctic Circle and the South Pole receive the sun's rays only for certain parts of the year. At the poles themselves there are six months of darkness and six months of sunshine. This, naturally, makes them the coldest

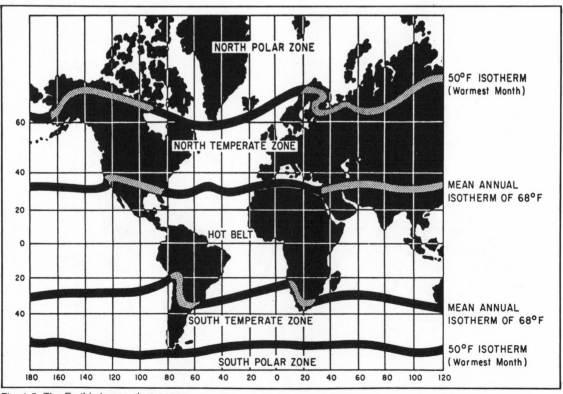

Fig. 1-5. The Earth's temperature zones.

zones. They are therefore known as the *Frigid* or *Polar Zones*.

As you can see, weather is greatly dependent on the relationship of the sun and the Earth. It also depends on solar radiation and how much falls on the Earth.

SOLAR RADIATION

Solar radiation is the total electromagnetic energy given off by the sun. The sun's surface emits gamma rays, X-rays, ultraviolet, visible light, infrared, heat, and electric waves. Even though the sun radiates in all wavelengths, about half the radiation is visible light and most of the remainder is infrared.

Insolation (an acronym for *in*coming *so*lar radi*ation*) is the rate at which solar radiation is received by a surface. There is a wide variety of differences in the amounts of radiation received over the various portions of the earth's surface, as you have just seen. These differences in heating are important in their effect on the weather.

Reflection occurs where a surface turns back a portion of the incident radiation toward its source. Some of the sun's radiation bounces off the Earth's atmosphere and back into space—about 40 percent. Earth surfaces also reflect radiation back through the atmosphere, each at different rates. Clouds return about 55 percent, new snow over 80 percent, land from 5 to 30 percent, water 10 percent or less.

The Earth's surface absorbs an average of about 50 percent of the incoming sun's rays. They warm the Earth and the air through which they pass—the atmosphere—and make the heating and cooling of the Earth uneven. This causes the weather we experience (Figs. 1-6 through 1-8).

LAYERS OF THE ATMOSPHERE

The Earth is surrounded by layers of gases with the densest layer—the *troposphere*—beginning at the surface and progressively thinning to the *exosphere* that borders space where there are no gases (Fig. 1-9).

Troposphere

The troposphere is the layer of air enveloping the Earth immediately above its surface. It is approximately 5½ miles (29,000 feet) thick over the poles, about 7½ miles (40,000 feet) thick in the mid-latitudes (such as over the United States) and about 11½ miles (61,000 feet) thick over the equator. These are average figures that change somewhat from day to day and from season to season. The troposphere is thicker in summer than in winter and during the day than during the night. All weather, as we know it, occurs in the troposphere.

The troposphere is composed of a mixture of several different gases. By volume, the composition of dry air in the troposphere is 78 percent nitrogen, 21 percent oxygen, nearly 1 percent argon, 0.03 percent carbon dioxide, and traces of helium, hydrogen, neon, krypton, and others (Fig. 1-10).

The air in the troposphere also contains a variable amount of water vapor. The maximum amount of water vapor that the air can hold depends on the temperature of the air and its pressure. The higher the temperature, the more water vapor it can hold at a given pressure.

The air also contains variable amounts of impurities such as dust, salt particles, soot, and chemicals. These impurities in the air are important because of their effect on visibility and especially because of the part they play in the condensation of water vapor. If the air were absolutely pure, there would be little condensation. These minute particles act as nuclei for the condensation of water vapor.

The temperature in the troposphere usually decreases with height, but there may be inversions for relatively thin layers at any level.

The *tropopause* is a transition layer between the troposphere and the stratosphere. It is not uniformly thick and is not continuous from the equator to the poles. The tropopause is characterized by little or no increase or decrease of temperature with increasing altitude. For most purposes it is considered one with the troposphere.

Stratosphere

The stratosphere directly overlies the tropopause and extends to about 30 miles (160,000

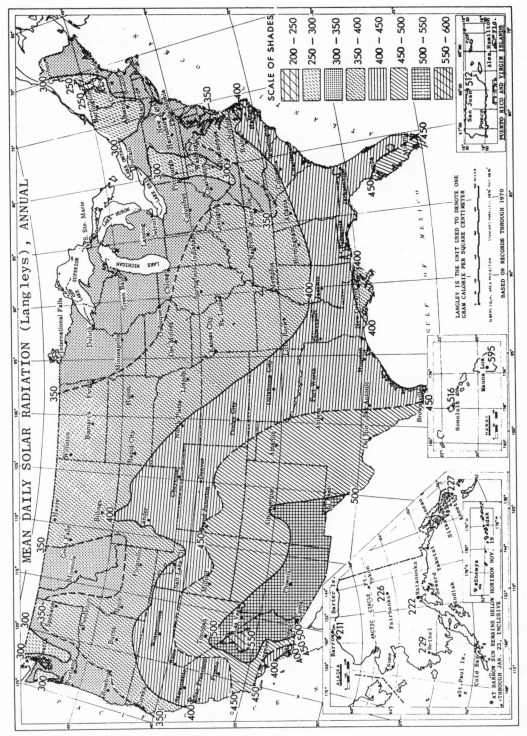

Fig. 1-6. Mean daily solar radiation, annually, in Langleys.

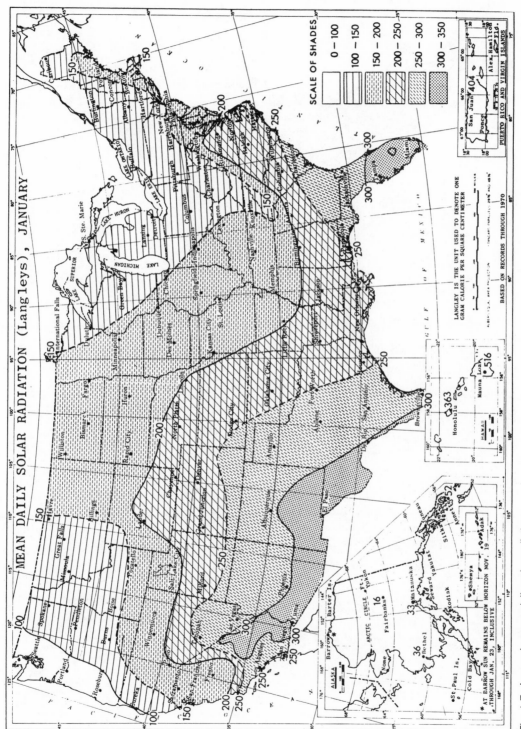

Fig. 1-7. January's mean daily solar radiation.

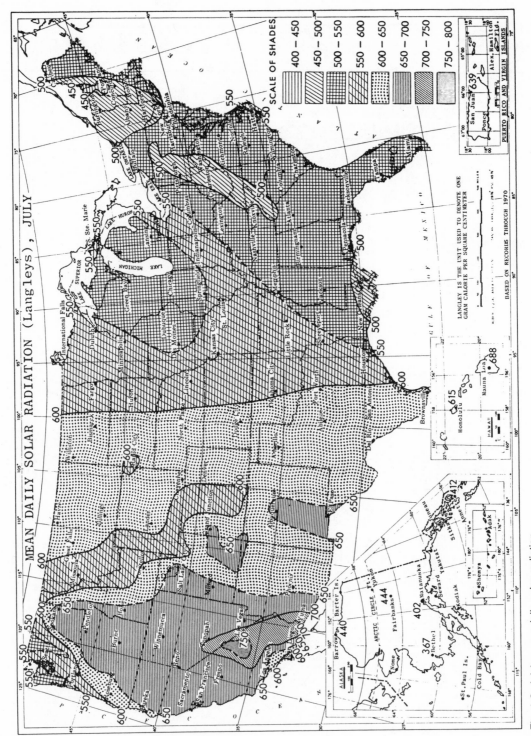

Fig. 1-8. July's mean daily solar radiation.

12

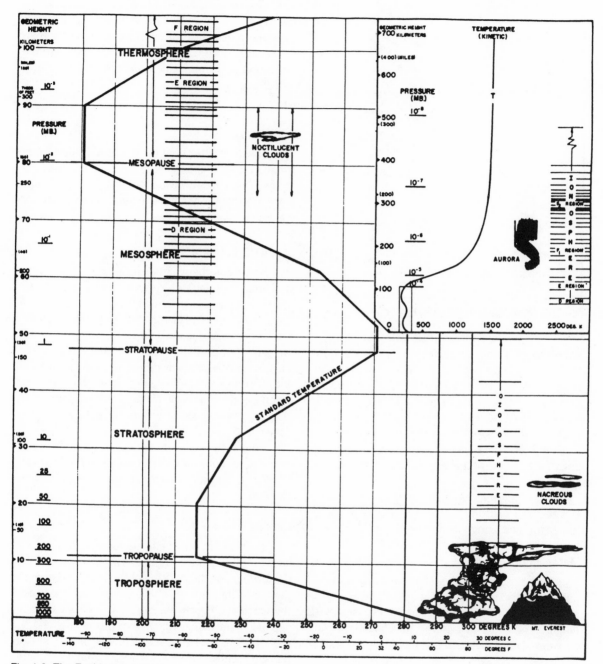

Fig. 1-9. The Earth's atmosphere.

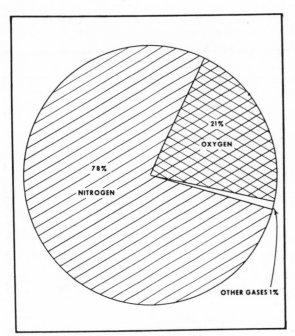

Fig. 1-10. Composition of the Earth's atmosphere.

feet). Temperature varies little with height in the stratosphere through the first 30,000 feet; however, in the upper portion the temperature increases approximately linearly to values nearly equal to surface temperatures. This increase in temperature through the stratosphere is attributed to the presence of *ozone*, which absorbs most of the incoming ultraviolet radiation. In fact, in some applications the stratosphere is called the *ozonosphere*.

The *stratopause* is the top of the stratosphere. It is the zone marking another reversal of temperature with increasing altitude (Fig. 1-11).

Mesosphere

The mesosphere is a layer approximately 20 miles (100,000 feet) thick that directly overlies the stratopause. The temperature decreases with height. The *mesopause* is the thin boundary zone between the mesosphere and the thermosphere. It is marked by a reversal of temperature; that is, temperature again increases with altitude. The transitions between layers are very gradual.

Thermosphere

The thermosphere, a second region in which the temperature *increases* with height, extends from the mesopause to outer space. The thermosphere was once called the *ionosphere* because the ionization of air molecules through this zone provides conditions that are favorable for radio propagation.

Exosphere

The very outer limit of the Earth's atmosphere is called the exosphere. It is the zone in which gas atoms are so widely spaced that they rarely collide with one another and have individual orbits around the Earth. There is no weather in the exosphere.

CLIMATIC CONTROLS

So why is the weather in New York different than that in Boise? The variation of climatic elements from place to place and from season to season is caused by several factors called *climatic controls*. The same basic factors that cause weather in the atmosphere also determine the climate of an area. These controls, acting in different combinations and with varying intensities, act upon temperature, precipitation, humidity, air pressure, and winds to produce many types of weather and therefore climate.

Four factors largely determine the climate of every region on Earth. They are latitude, land and water distribution, topography, and the prevailing elements.

Latitude

Perhaps no other climatic control has such a marked effect on the weather and climate as does the *latitude*, or the position of the Earth relative to the sun. The angle at which rays of sunlight reach the earth and the number of "sun" hours each day depend on the distance from the equator (Fig. 1-12).

Regions under direct or nearly direct rays of the sun receive more heat than those under oblique rays. The heat brought about by the slanting rays of early morning may be compared with the heat that is caused by the slanting rays of winter. The heat which is due to the more nearly direct rays of the midday sun may be compared to the heat resulting from the summer sun.

The length of the day, like the angle of the sun's rays, influences the temperature. The length of the day varies with the latitude and the season of the year (Fig. 1-13).

The hot and humid climates of equatorial Africa and South America are good examples of the influence that latitude has on climate. At no time during the year are the sun's rays at much of an oblique angle. Therefore, there is little difference between the mean temperatures for the coldest and warmest month. Contrast this picture with the opposite extreme in the far north, where the sun is

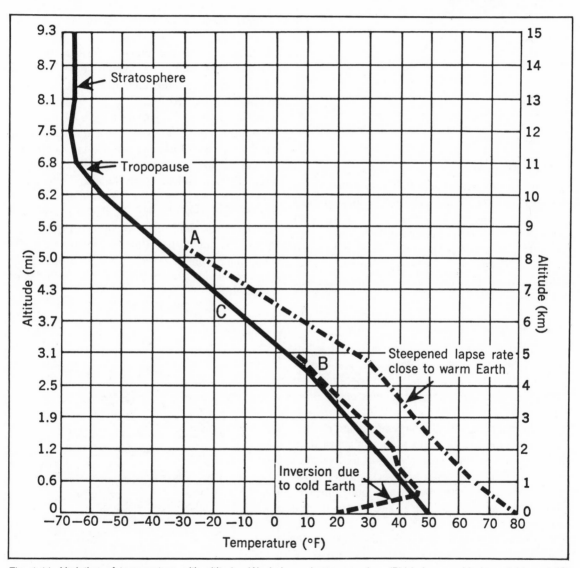

Fig. 1-11. Variation of temperature with altitude: (A) during a hot sunny day, (B) during a cold clear night, and (C) middle-latitude average.

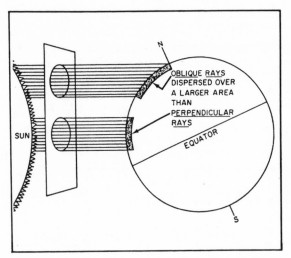

Fig. 1-12. Different areas of the earth receive different amounts of sunshine.

below the horizon for a great deal of the time during the winter, producing cold temperature that breed powerful polar cold fronts. During the long hours of summer daylight, the sun's rays make such a small angle with the Earth's surface that the energy received in any given polar area is extremely small and the sun's effectiveness is minimized. It is, however, sufficient to thaw lakes and weaken the polar air masses (Fig. 1-14).

Land and Water Distribution

Because land and water heat and cool at different rates, the location of continents and oceans greatly alters the pattern of air temperature and influences the sources and direction of air mass movement.

Coastal areas take on the temperature characteristics of the land or water to their windward. In latitudes of prevailing westerly winds, for example, west coasts of continents have oceanic temperatures and east coasts have continental temperatures. The temperatures are determined by the windflow.

Since the upper layers of the ocean are nearly always in a state of violent stirring, heat losses or heat gains occurring at the sea surface are distributed throughout a large volume of water. This mixing process sharply reduces the temperature contrasts between day and night, and between winter and summer over oceanic areas.

Land surfaces are not subject to such a mixing process. Thus, violent contrasts between seasons and between day and night are created in the interiors of continents. During winter, a large part of the sparse insolation is reflected back toward space by the snow cover that extends over large portions of the northern continents. For these reasons, the northern areas serve as manufacturing plants for dry polar air.

The great temperature difference between land and water surfaces, which reverses between the two seasons, determines to a great extent the seasonal weather patterns.

The nature of the surface affects the local heat distribution, too. Color, texture, and vegetation influence the rate of heating and cooling. Generally, dry surfaces heat and cool faster than moist surfaces. Plowed fields, sandy beaches, and paved roads become hotter than surrounding meadows and wooded areas. During the day, air is warmer over a plowed field than over a forest or swamp. During the night the situation is reversed. The small plane pilot is critically aware of this as he fights for control of his plane at low altitudes on warm days.

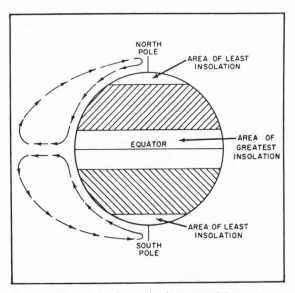

Fig. 1-13. How air begins to circulate around the earth.

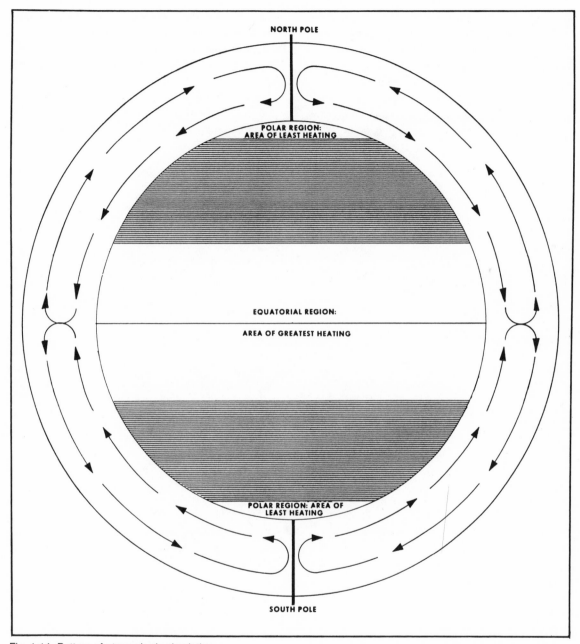

Fig. 1-14. Pattern of atmospheric circulation.

Topography

Over land, climates may vary radically within a short distance because of land forms and variations of elevation. The height of an area above sea level exerts a considerable influence on its climate. For instance, a place located on the equator in the high Andes of South America would have a climate quite different from a place located a few feet above sea level at the same latitude.

Mountainous terrain has a powerful influence

on climates, especially the long high chains of mountains (such as the Rockies) that act as climatic dividers. These obstacles deflect the tracks of storms and block the passage of air masses on lower levels. If the mass is strong enough to go over a mountain, it will be greatly modified by the time it reaches the other side. East-west mountain chains, such as the Alps, modify south-running polar air and make regions to the south warmer than other points at the same latitude—as is the case in Italy.

Prevailing Elements

Each region of the earth has its own prevailing elements—air masses, fronts, temperature, wind, and precipitation—that play a part in making weather. These will be covered in the next two chapters to help you understand weather.

Chapter 2

Air Masses

The air mass concept is one of the most important developments in the history of meteorology and a major key to modern weather analysis and forecasting. An air mass is a large body of air whose physical properties, particularly temperature and moisture distribution, are the same on all levels. Forecasting is largely a matter of recognizing the various air masses in the weather picture, determining their characteristics, and predicting their behavior.

AIR MASS CLASSIFICATIONS

Air masses have been classified geographically and thermodynamically. The *geographical classification*, which refers to the source region of the air mass, divides air masses into four basic categories. These are arctic/antarctic (A), polar (P), tropical (T), and equatorial (E). Air masses are further divided into maritime or ocean (M) and continental or land (C).

An air mass is considered *maritime* if its source of origin is over an oceanic surface. If the air mass originates over a land surface, it is considered *con-tinental*. Maritime arctic/antarctic air masses are rare, since there is a predominance of land masses or icefields in the polar regions. Virtually all equatorial air masses are considered to be maritime in origin.

There are two less common air masses that should be mentioned. The superior (S) air mass is generally found over the southwestern United States and the monsoon (M) air mass is a local condition in India and southeastern Asia (Figs. 2-1 and 2-2).

The types of air masses can easily be remembered by using the letters in the word tape to stand for the first letter of each basic type of air mass: tropical, arctic/antarctic, polar, and equatorial.

The *thermodynamic classification* applies to the relative warmth or coldness of the air mass. A warm air mass (w) is one that is warmer than the underlying surface. A cold air mass (k) is one that is colder than the underlying surface. For example, a continental polar cold air mass is classified or written as *cPk*. An *mTw* classification indicates a maritime tropical warm air mass.

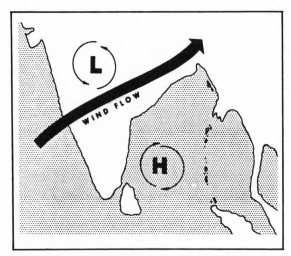

Fig. 2-1. Pressure system movement causes summer monsoons.

Air masses can usually be identified by the type of clouds within them. Cold air masses usually show *cumuliform* clouds while warm air masses contain *stratiform* clouds. (More on this later.)

Sometimes, and with some air masses, the thermodynamic classification may change from night to day. A particular air mass may show *k* characteristics during the day and *w* characteristics at night and vice versa.

SOURCE REGIONS

The air mass *source region* is the area where the air mass originates. The ideal condition for the production of an air mass is the stagnation of air over a rather uniform surface (water, land, or icecap) of uniform temperature and humidity. The length of time an air mass stagnates over its source region depends on the surrounding pressures. The air acquires definite properties and characteristics from the surface up and becomes virtually *homogeneous* throughout—its properties become rather uniform at each level.

The source regions for the world's air masses are shown in Fig. 2-3. Note the uniformity of the underlying surfaces. Also note the relatively uniform climatic conditions in the various source regions such as the southern North Atlantic and Pacific Oceans for maritime tropical air and the

deep interiors of North America and Asia for continental polar air.

Arctic air is a permanent high-pressure area in the vicinity of the North Pole, within which is found the arctic air mass source region. In this region there is a gentle flow of air over the polar icefields, allowing the formation of the arctic air mass. The air is characterized by being dry aloft and very cold and stable in the lower altitudes.

Antarctic air is developed in the antarctic region. It is colder at the surface and other levels than arctic air in fall and winter.

Continental polar air source regions consist of all the land areas dominated by the Canadian and Siberian high-pressure cells (Fig. 2-4). In the winter these regions are covered by snow and ice. Because of the intense cold and the absence of water bodies, very little moisture is taken into the air in these regions. Note that the word "polar" when applied to air mass designations does not mean air at the poles. Polar air is generally found between 40 and 60 degrees latitude and, except for ground over northern and central Asia, is generally drier than arctic air.

Maritime polar air source regions consist of the open unfrozen polar sea areas in the vicinity of 60 degrees latitude, north and south. Such areas are sources of moisture for polar air masses. Consequently, air masses forming over these regions

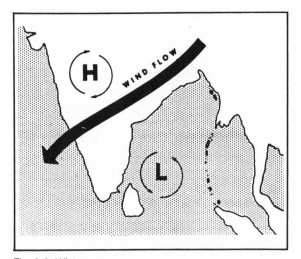

Fig. 2-2. Winter monsoons.

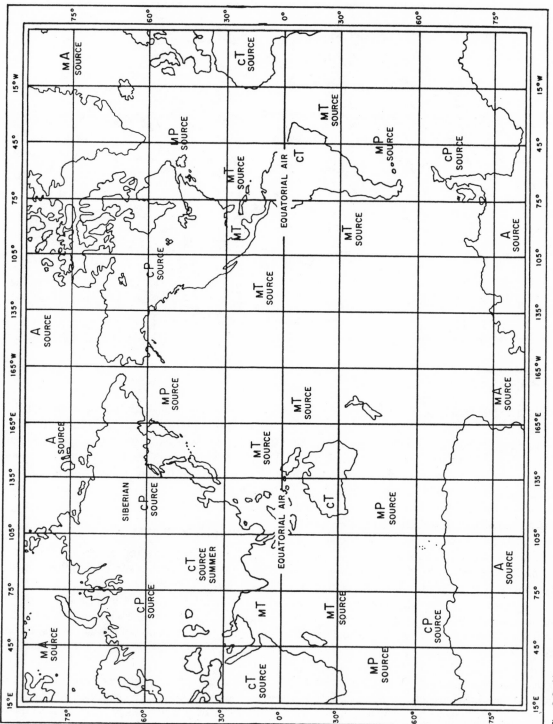

Fig. 2-3. Air mass source regions.

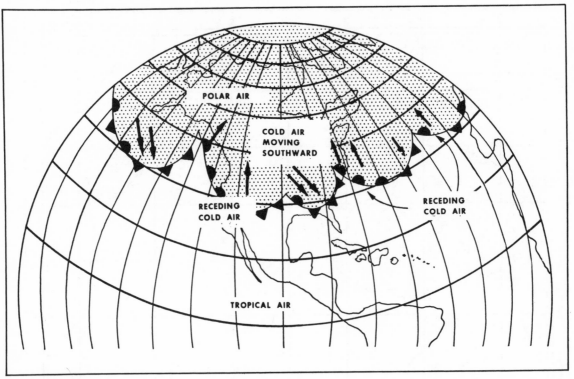

POLAR AIR

COLD AIR MOVING SOUTHWARD

RECEDING COLD AIR

RECEDING COLD AIR

TROPICAL AIR

Fig. 2-4. Polar front.

are moist, but the moisture is sharply limited by the temperature.

Continental tropical air source regions can be any significant land areas lying in the tropical regions, generally between 25 degrees north latitude and 25 degrees south latitude. The large land areas found there are usually desert regions, such as the Sahara and Kalahari Deserts of Africa, the Arabian Desert, and the interior of Australia. The air over these land areas is hot and dry.

Maritime tropical air source regions are the large zones of tropical sea along the belt of the subtropical *anticyclones* or high pressure cells that stagnate in this area most of the year. The air is warm due to low latitude and is able to hold considerable moisture.

Equatorial air's source region is the area from about 10 degrees north to 10 degrees south latitudes within which the thermal equator is found. It is essentially an oceanic belt which is very warm and has a high moisture content. Convergence of

the trade winds from both hemispheres and the intense isolation over this region causes lifting of the air, which is unstable and moist, to high levels (Fig. 2-5). The weather associated with these conditions is characterized by thunderstorms throughout the year.

HOW AIR MASSES CHANGE

When an air mass moves out of its source region, there are a number of factors which act upon it to change its properties (Fig. 2-6). These modifying influences do not occur separately. For instance, in the passage of cold air over warmer water surfaces, there is not only a release of heat to the air, but also some moisture in the form of rain (Fig. 2-7).

As an air mass expands and slowly moves out of its source region, it travels along a certain path. The surface over which this path takes the air mass after leaving its source modifies the air mass. The

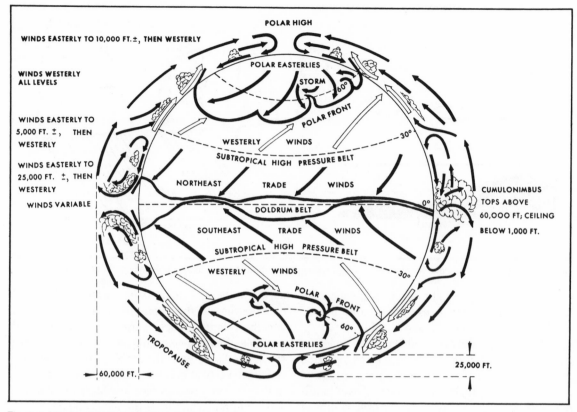

Fig. 2-5. Idealized pattern of atmospheric circulation.

way it rotates, clockwise or counterclockwise, also has a bearing on its modification. The time the air mass has been out of its source region will determine to a great extent the characteristics of the air mass.

Surface Conditions

The first modifying factor on an air mass as it leaves its source region is the type and condition of the surface over which it travels. Here, the factors of surface temperature, moisture, and topography must be considered (Fig. 2-8).

The temperature of the surface relative to that of the air mass will modify not only the temperature, but its stability as well. For example, if the air mass is warm and moves over a colder surface, such as a tropical air moving over colder water, the cold surface cools the lower layers of the air mass and its stability is increased. This stability will extend to

the upper layers in time and condensation in the form of fog or low stratus normally occurs. If the air mass moves over a surface that is warmer, such as polar continental air moving out from the continent in winter over warmer water, the warm water heats the lower layers of the air mass, increasing instability that consequently spreads to higher layers. The changes in stability of the air mass give valuable indications of the cloud types that will form as

Fig. 2-6. Air masses vary in water vapor amounts and dew point.

23

COOL CONTINENT

WARM OCEAN

Fig. 2-7. When continental polar air of winter moves from cool continent to warm ocean, it usually produces precipitation.

well as the type of precipitation. Also, the increase or decrease in stability gives further indications of the lower layer turbulence and visibility.

The air mass may be modified in its moisture content by the addition of moisture by evaporation or by the removal of moisture by condensation and precipitation. If the air mass is moving over continental regions, the existence of unfrozen bodies of water can greatly modify the air mass and, in the case of an air mass moving from a continent to an ocean, the modification can be considerable. In general, depending upon the temperature of the two

surfaces, the movement over a water surface will *increase* the moisture content of the lower layers and the relative temperature of the surface. For example, the passage of cold air over a warm water surface will decrease the stability of the air with resultant vertical currents. The passage of warm moist air over a cold surface increases the stability and could result in fog as the air is cooled.

The effect of topography is evident primarily in mountainous regions. The air mass is modified on the windward side (the side from which the wind blows) by the removal of moisture through precipi-

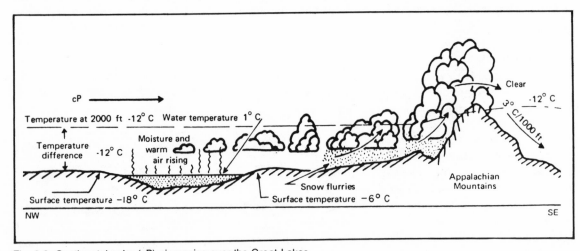

Fig. 2-8. Continental polar (cP) air moving over the Great Lakes.

tation with a decrease in stability. As the air descends on the other side of the mountain, the stability increases as the air becomes warmer and drier (Fig. 2-9).

Path

After an air mass has left its source region, the path it follows, and its turning clockwise or counterclockwise, have a great effect on its stability. If the air follows a *cyclonic trajectory* (counterclockwise in the Northern Hemisphere and clockwise in the Southern Hemisphere), its stability in the upper levels is decreased. On the other hand, if the trajectory is *anticyclonic* (opposite directions), its stability in the upper levels is increased.

Age

Although the age of an air mass in itself cannot modify the air, it will determine to a great extent the amount of modification that takes place. For example, an air mass that has recently moved from its source region will not have had time to become modified significantly. However, an air mass that has moved into a new region and stagnated for some time, and is now old, will be found to have lost many of its original characteristics.

In Table 2-1 the two modifying influences are classified *thermal* and *mechanical*. The table indicates the modifying process, what takes place, and the resultant change in stability of the air mass. Remember, these processes do not occur independently, but two or more processes are usually in evidence at the same time. Also keep in mind that conditions indicated are only *average* conditions and that each individual case may be quite different. However, understanding the types, sources, and modifying factors in air masses will give you a greater understanding of weather and how to forecast it accurately.

UNDERSTANDING FRONTS

In weather, the term *front* means a boundary separating two different air masses. From this definition, the close relationship that exists between air masses and fronts can be readily seen. In fact, without the air masses, there would be no fronts.

The centers of action bring together air masses of different physical properties. The region of transition between two air masses is called a *frontal zone*. The primary frontal zones of the Northern Hemisphere are the arctic frontal zone and the polar frontal zone. The most important frontal zone affecting the United States is the polar

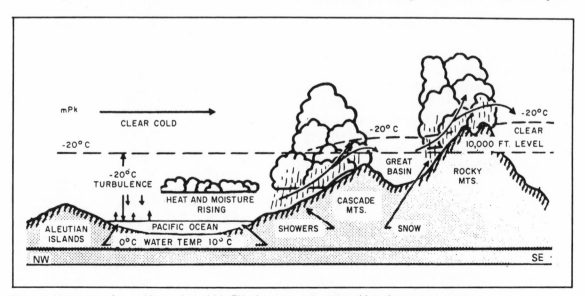

Fig. 2-9. Movement of a maritime polar cold (mPk) air mass southeastward into the western states.

Table 2-1. Thermal and Mechanical Air Mass Modifications.

The Process	What Takes Place	Results
A. THERMAL		
1. Heating from below.	Air mass passes from over a cold surface to a warm surface, or surface under air mass is heated by sun.	Decrease in stability.
2. Cooling from below.	Air mass passes from over a warm surface to a cold surface, OR radiational cooling of surface under air mass takes place.	Increase in stability.
3. Addition of moisture.	By evaporation from water, ice, or snow surfaces, or moist ground, or from rain-drops or other precipitation which falls from overrunning saturated air currents.	Decrease in stability.
4. Removal of moisture.	By condensation and precipitation from the air mass.	Increase in stability.
B. MECHANICAL		
1. Turbulent mixing.	Up- and down-draft.	Tends to result in a thorough mixing of the air through the layer where the turbulence exists.
2. Sinking.	Movement down from above colder air masses or descent from high elevations to lowlands, subsidence and lateral spreading.	Increases stability.
3. Lifting.	Movement up over colder air masses or over elevations of land or to compensate for air at the same level converging.	Decreases stability.

front. The polar front is a region of transition between the cold polar air and warm tropical air. During the winter months, the polar front pushes farther southward (due to the greater density of the polar air) than during the summer months. In the summer months, the front seldom moves farther south than the central United States.

On a surface map, a front is indicated by a line separating two air masses. This is only a picture of the surface conditions. These air masses also have vertical extent (Fig. 2-10).

A cold air mass, being heavier, tends to underrun a warm air mass. Thus, the cold air is below and the warm air is above the surface. The slope of a frontal surface is usually between 1:50 (1 mile vertical for 50 miles) horizontal) for a cold front and 1:300 (1 mile vertical for 300 miles horizontal) for a warm front. For example, 100 miles from the place where the frontal surface meets the ground, the frontal surface might be somewhere between 2000 feet and 2 miles above the Earth's surface, depending on the slope (Fig. 2-11). The slope of a front is of

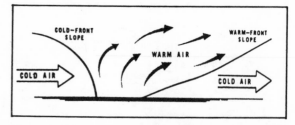

Fig. 2-10. Vertical view of a frontal system (without clouds shown).

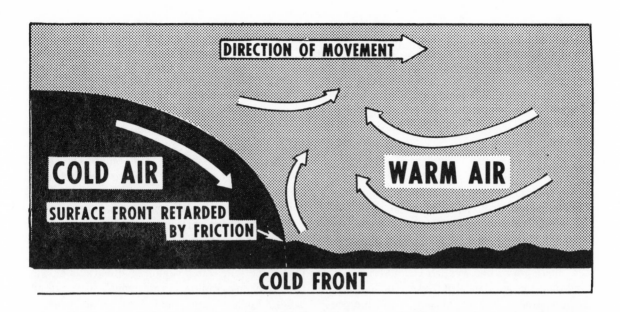

COLD FRONT

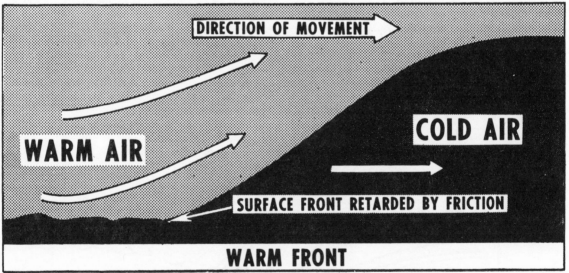

WARM FRONT

The slopes in the above fronts are exaggerated for diagrammatic purposes.

The scale below represents a true vertical scale of 1 to 100.

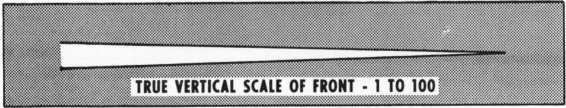

TRUE VERTICAL SCALE OF FRONT - 1 TO 100

Fig. 2-11. Vertical cross section of cold and warm fronts showing frontal slopes.

considerable importance in visualizing and understanding the weather along the front.

Note being greatly affected by daily heating and cooling of the Earth's surface, the dew point is normally more consistent than the temperature through the day, except with the passage of a front. Therefore, the dew point is a good index of frontal passage.

COLD FRONTS

A *cold front* is the line of discontinuity along which a wedge of cold air is underrunning and displacing a warmer air mass. This term is also used, but inexactly, when referring to a cold frontal surface.

There are certain weather characteristics and conditions that are typical of cold fronts. In general, the temperature and humidity *decrease*, the pressure *rises* and, in the Northern Hemisphere, the wind *shifts* (usually from southwest to northwest) with the passage of a cold front. The distribution of precipitation depends primarily on the vertical velocities in the warm air mass. On the basis of this

latter factor, cold fronts are classified as slow-moving and fast-moving cold fronts.

Slow-Moving Cold Fronts

With the slow-moving cold front there is a general upward motion of warm air along the entire frontal surface except for pronounced lifting along the lower portion of the front. The average slope of the front is approximately 1:100 miles. The cloud and precipitation area is extensive and is characterized by cumulonimbus and nimbostratus clouds, and showers and thunderstorms at and immediately to the rear of the surface front (Chapter 3 covers clouds and how to recognize them by type). This area is followed by a region of rain and nimbostratus clouds merging into a region of altostratus clouds and then cirrostratus clouds, which may extend several hundred miles behind the surface front.

The development of cumulonimbus clouds, showers, and thunderstorms is largely dependent on the original instability characteristics of the *warm* air mass. Within the cold air mass there may be some stratified clouds in the rain area, but there

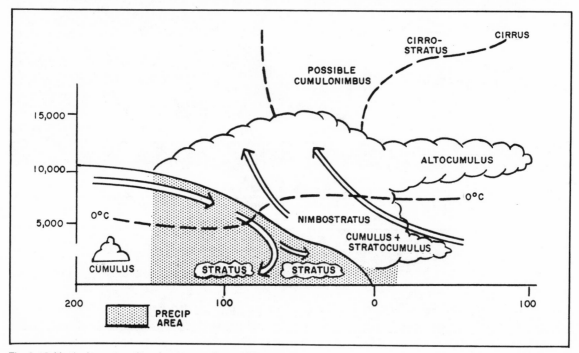

Fig. 2-12. Vertical cross section of a slow-moving cold front.

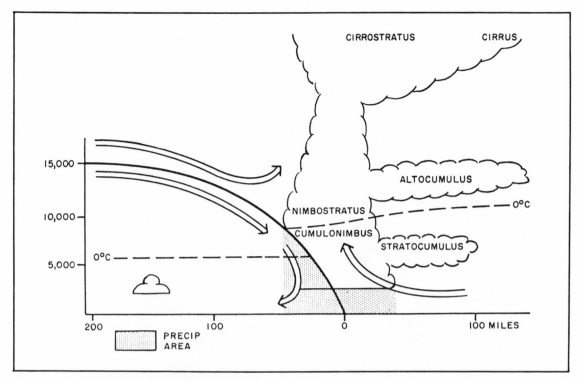

Fig. 2-13. Vertical cross section of a fast-moving cold front.

are no clouds in the cold air beyond this area unless the cold air mass is unstable. In the latter case, some cumulus clouds may develop. This type of front is slow moving; 15 knots (17 miles an hour) may be considered average. Figure 2-12 shows a cross section of a typical slow-moving cold front.

Fast-Moving Cold Fronts

With the fast-moving cold front the high level warm air along the frontal surface descends and the warm air near the surface is pushed vigorously upward. This type of front has a slope of 1:40 to 1:80 miles and usually moves rapidly; 25 to 30 knots (29 to 34 miles an hour) may be considered an average speed of movement. Because of these factors, there is a relatively narrow but often violent band of weather associated with the passing of a fast-moving cold front. If the warm air mass is conditionally unstable and moist, cumulonimbus clouds, showers, and thunderstorms occur just ahead of and at the surface front, and rapid clearing occurs be-

hind the front. Frequently, altostratus and altocumulus cloud layers form and drift ahead of the main cloud bank. The more unstable the warm air mass, the more violent the weather. If the warm air is relatively dry, this type of front may not produce precipitation or clouds. It is with the fast-moving cold front that squall lines (violent winds and precipitation) are associated.

Figure 2-13 shows a typical cross section of a fast-moving cold front. It also shows the cloud shield, precipitation shield and frontal slope (exaggerated in the vertical) associated with this type of front.

WARM FRONTS

A warm front is the line of discontinuity between the forward edge of an advancing mass of relatively warm air and a retreating, relatively colder air mass. As in the case of the cold front, this term is used inexactly when referring to a warm frontal surface.

29

Certain characteristics and weather conditions are associated with warm fronts. The winds shift from southeast to southwest or west, but the shift is not as pronounced as with the cold front. The temperatures are colder ahead of the front and warmer after the passage of the front. The average slope of a warm front is 1:150.

A characteristic phenomenon of a typical warm front is the sequence of cloud formations. They are noticeable in the following order: cirrus, cirrostratus, altostratus, nimbostratus, and stratus. The cirrus clouds may appear 700 to 1000 miles ahead of the surface front followed by cirrostratus about 600 miles and altostratus about 500 miles ahead of the surface front.

Precipitation in the form of continuous or intermittent rain, snow, or drizzle is frequent as much as 300 miles in advance of the surface front. Surface

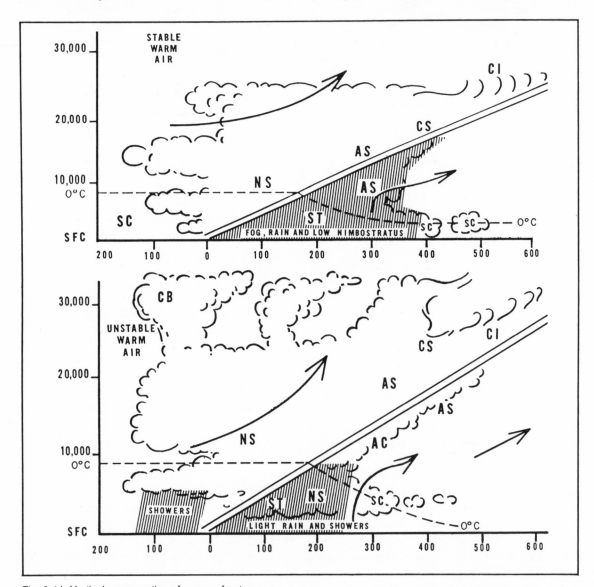

Fig. 2-14. Vertical cross section of a warm front.

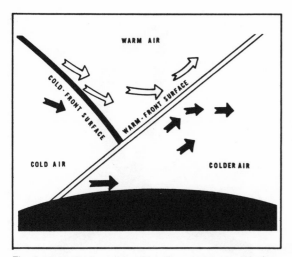

Fig. 2-15. Vertical cross section of a warm type occlusion.

precipitation is associated with the nimbostratus in the warm air above the frontal surface, and from stratus in the cold air. However, when the warm air is convectively unstable, showers and thunderstorms may occur in addition to the steady precipitation. Fog is common in the cold air ahead of a warm front.

Clearing usually occurs after the passage of a warm front, but under some conditions drizzle and fog may occur within the warm sector. Warm fronts usually move in the direction of the isobars of the warm sector; in the Northern Hemisphere this is usually east to northeast. Their speed of movement is normally less than that of cold fronts. On the average it may be considered to be about 10 knots (12 miles an hour).

Figure 2-14 summarizes pictorially the main features of warm fronts under average conditions.

OCCLUDED FRONTS

An occluded front occurs when a cold front overtakes a warm front. One of the two fronts is lifted aloft and the warm air between the fronts is shut off from the Earth's surface. An occluded front is often referred to as an *occlusion* (a closing). The type of occlusion is determined by the temperature difference between the cold air in advance of the warm front and the cold air behind the cold front.

Figure 2-15 shows a warm-type occlusion. If the air in advance of the warm front is colder than the air behind the cold front, the cold front rides up the warm frontal slope.

If the cold air ahead of the warm front is warmer than the cold air behind the cold front, the cold frontal surface underruns the warm front and the occluded front is called a cold-type occlusion, shown in Fig. 2-16.

The primary difference between a warm-type and cold-type occlusion is the location of the associated upper front in relation to the surface front (Fig. 2-17). In a warm-type occlusion, the upper cold front *precedes* the surface occluded front by as much as 200 miles. In the cold-type occlusion, the upper warm front *follows* the surface occluded front by 20 to 50 miles.

Since the occluded front is a combination of a cold front and a warm front, the resulting weather is that of the cold front's narrow band of violent weather and the warm front's wide-spread area of cloudiness and precipitation, all occurring in combination along the occluded front. The most violent weather occurs at the tip of the occlusion—the point at which the cold front is overtaking the warm front.

STATIONARY FRONTS

When a front is stationary, the cold air mass, as a whole, does not move either toward or away from

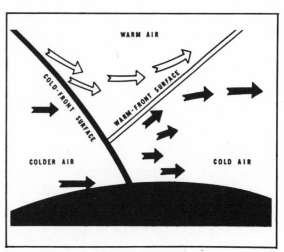

Fig. 2-16. Vertical cross section of a cold type occlusion.

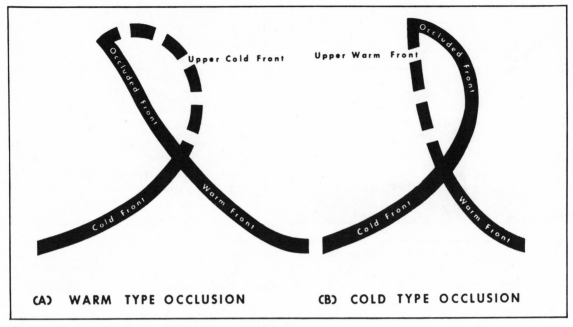

Fig. 2-17. Occlusions (in the horizontal) and associated upper fronts.

the front. In terms of wind direction, this means that the wind above the friction layer blows neither toward nor away from the front, but parallel to it. It follows that the isobars (lines of identical barometric pressure—see Chapter 3) are also nearly parallel to a stationary front. This characteristic makes it easy to recognize a stationary front on a weather map.

The frictional inflow of warm air toward a stationary front causes a slow upglide of air on the frontal surface. As the air is lifted to and beyond its lifting condensation level, clouds form in the warm air above the front.

If the warm air in a stationary front is stable, the clouds are stratiform. Drizzle may then fall. As the air is lifted beyond the freezing level, icing

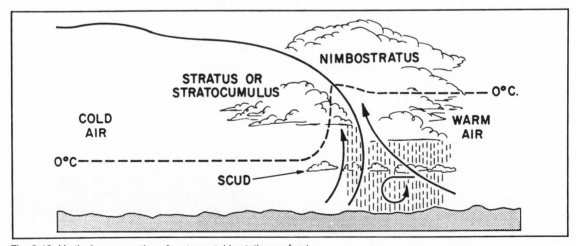

Fig. 2-18. Vertical cross section of a steep stable stationary front.

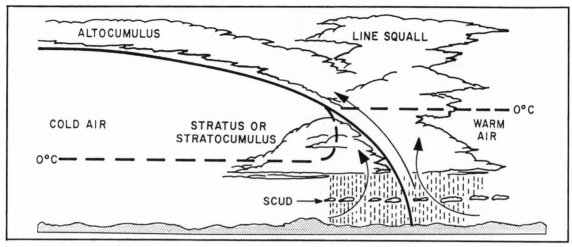

Fig. 2-19. Vertical cross section of a steep unstable stationary front.

conditions develop and light rain or snow may fall. At very high levels above the front ice clouds are present (Fig. 2-18).

If the warm air is conditionally unstable and sufficient lifting occurs, the clouds are then cumuliform or stratiform with cumuliform sections. If the energy release is great (warm, moist, unstable air), thunderstorms result. Rainfall is generally showery (Figs. 2-19, 2-20).

Within the cold air mass extensive fog and low ceiling may result. The cold air is saturated by warm rain or drizzle falling through it from the warm air mass above. If the temperature is below 0° C (32° F), icing may occur, but generally it is light (Fig. 2-21).

The width of the band of precipitation and low ceiling varies from 50 miles to about 200 miles, depending on the slope of the front and the tempera-

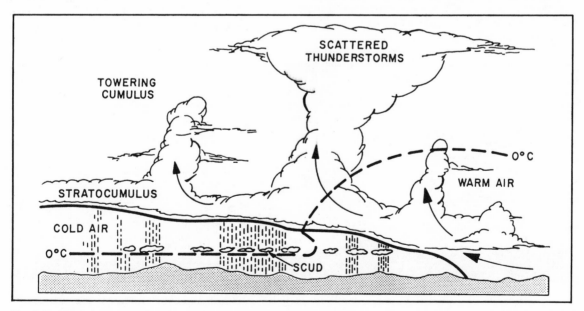

Fig. 2-20. Vertical cross section of a shallow unstable stationary front.

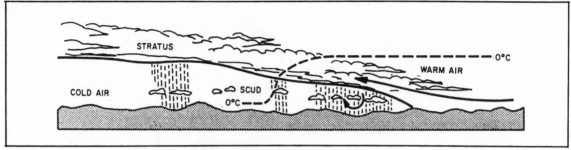

Fig. 2-21. Vertical cross section of a shallow stable stationary front.

tures of the air masses. One of the most annoying characteristics of a stationary front to airmen and travelers is that it may greatly hamper airport operations by persisting in the area for several days (Fig. 2-22).

LOWS AND HIGHS

These moving air masses of warmer and colder air, along with their fronts, cause changes in the barometric pressure. In fact, fronts can be found by drawing a connecting line on a weather map between points of low barometric pressure—called a *trough*.

A *low* is a pressure system in which the barometric pressure decreases toward the center and the windflow around the system is counterclockwise in the Northern Hemisphere (Fig. 2-23). The terms *low,* and *cyclone* or *wave cyclone*, are interchangeable (Fig. 2-24). Any pressure system in the Northern Hemisphere with a counterclockwise (cyclonic) windflow is a *cyclone*. The term "cyclone" is inaccurately used to describe some violent storms. Low pressure systems with severe storm characteristics are more accurately called hurricanes, typhoons, tropical storms, tornadoes, or waterspouts to identify the exact nature of the storm.

Low pressure systems usually bring cloudy weather with rain or snow, and move through quickly. Some lows, called *tropical lows*, bring warm weather. *Thermal lows* are caused by intense surface heating and resulting low air density over barren continental areas. They are relatively dry with few clouds and practically no precipitation. These thermal lows are almost stationary and predominate over continental areas in the summer.

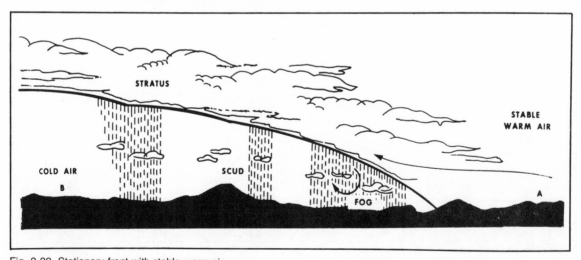

Fig. 2-22. Stationary front with stable warm air.

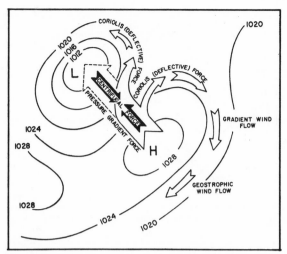

Fig. 2-23. Examples of circulation about high and low pressure systems.

A *high* is a pressure system in which the barometric pressure increases toward the center and the windflow around the system is clockwise in the Northern Hemisphere (Fig. 2-25). The terms *high* and *anticyclone* (opposite of cyclone) are interchangeable. Any pressure system in the Northern Hemisphere with a clockwise (anticyclonic) windflow is an anticyclone or high (Fig. 2-26).

High pressure systems predominate over cold surfaces where the air is dense. They are more intense over continental areas in winter and oceanic areas in summer. They usually bring fair weather and stay awhile, offering clearer skies, less winds, and better visibility (Fig. 2-27).

In future discussions of highs and lows you will hear three other terms: ridge, trough, and col. A *col* is a saddleback region between two highs or two

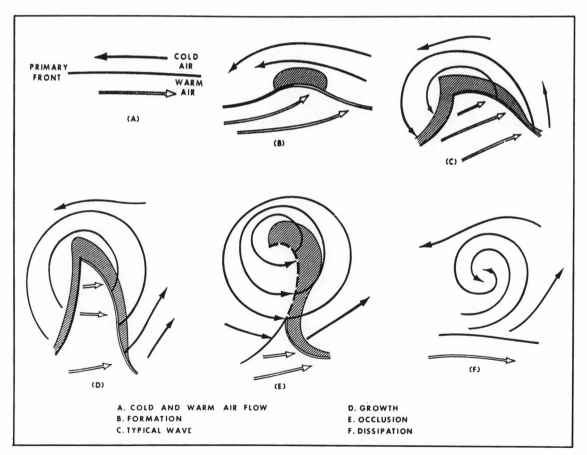

Fig. 2-24. Life cycle of a wave cyclone.

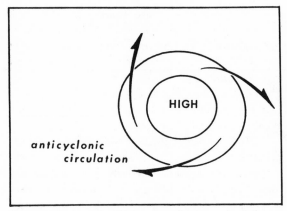

Fig. 2-25. Anticyclone (high pressure) system circulation.

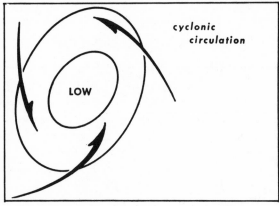

Fig. 2-26. Cyclone (low pressure) system circulation.

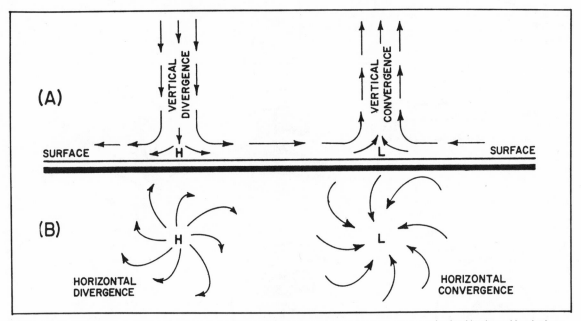

Fig. 2-27. Horizontal and vertical views at the movement of high and low pressure systems in the Northern Hemisphere.

lows. The weather is erratic and unpredictable. A *trough* is an elongated area of low pressure with the lowest pressure along the trough line, where weather is frequently violent. A *ridge* is an elongated area of high pressure with highest pressure along the ridge line. It usually offers good weather.

Air masses and fronts both make and reflect changes in the elements of the atmosphere. Learning more about these elements will improve your understanding of the weather.

Chapter 3

The Elements

What we call "the weather" can be broken down into six elements: temperature, pressure, wind, moisture, clouds, and precipitation. These are measurable forces in the earth's atmosphere that can be used to read and forecast weather.

These elements interact in a variety of combinations to give us clear and temperate spring days, sultry nights, Christmas snowstorms, and red sunsets.

TEMPERATURE

All substances are made up of super-small molecules, which are in more or less rapid motion. Temperature is a measure of the average speed of these molecules. As the velocity of these molecules increases in a substance under constant pressure, the temperature of the substance increases.

Surface air temperatures are usually measured with liquid-in-glass thermometers (to be discussed in Chapters 5 and 6 on basic and advanced instruments).

Two fixed temperatures—the melting point of ice and the boiling point of water (at standard pressure)—are used to calibrate thermometers. The two scales in common use are *centigrade* (or Celsius) and *Fahrenheit* (Fig. 3-1). The centigrade scale was devised by Anders Celsius during the 18th century. The terms *centigrade scale* and *degrees centigrade* have been used for many years, but the accepted terminology now is *Celsius scale* and *degrees Celsius*.

To convert from Celsius to Fahrenheit, use the equation:

$$F = 9/5°C + 32$$

To convert from Fahrenheit to Celsius, use the equation:

$$C = 5/9 \ (°F - 32)$$

Aviators know that there is normally an overall decrease of temperature in the troposphere as an aircraft gains altitude, because the air nearest the ground (which has been heated by incoming solar radiation) receives the largest amount of heat. The

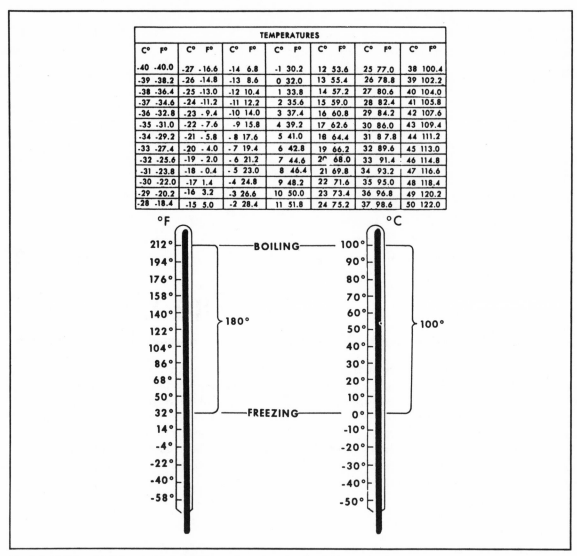

TEMPERATURES

C°	F°	C°	F°	C°	F°	C°	F°	C°	F°	C°	F°	C°	F°
-40	-40.0	-27	-16.6	-14	6.8	-1	30.2	12	53.6	25	77.0	38	100.4
-39	-38.2	-26	-14.8	-13	8.6	0	32.0	13	55.4	26	78.8	39	102.2
-38	-36.4	-25	-13.0	-12	10.4	1	33.8	14	57.2	27	80.6	40	104.0
-37	-34.6	-24	-11.2	-11	12.2	2	35.6	15	59.0	28	82.4	41	105.8
-36	-32.8	-23	-9.4	-10	14.0	3	37.4	16	60.8	29	84.2	42	107.6
-35	-31.0	-22	-7.6	-9	15.8	4	39.2	17	62.6	30	86.0	43	109.4
-34	-29.2	-21	-5.8	-8	17.6	5	41.0	18	64.4	31	87.8	44	111.2
-33	-27.4	-20	-4.0	-7	19.4	6	42.8	19	66.2	32	89.6	45	113.0
-32	-25.6	-19	-2.0	-6	21.2	7	44.6	20	68.0	33	91.4	46	114.8
-31	-23.8	-18	-0.4	-5	23.0	8	46.4	21	69.8	34	93.2	47	116.6
-30	-22.0	-17	1.4	-4	24.8	9	48.2	22	71.6	35	95.0	48	118.4
-29	-20.2	-16	3.2	-3	26.6	10	50.0	23	73.4	36	96.8	49	120.2
-28	-18.4	-15	5.0	-2	28.4	11	51.8	24	75.2	37	98.6	50	122.0

Fig. 3-1. Celsius and Fahrenheit temperature scale and conversion table.

variation in temperature with altitude is called the temperature *lapse rate* and is usually expressed in degrees per thousand feet. On one day, the air may have a decrease in temperature of 3°C for each thousand feet gained in altitude. Another day may show a decrease of 1°C per thousand feet. A third day may reveal the temperature *increasing* for a distance of one or two thousand feet above the ground (called an *inverted lapse rate* or *inversion*), and thereafter decreasing at the rate of 3°C per thousand feet. If such observations taken day after day over thousands of locations on the earth were averaged, the average lapse rate would be about 2°C decrease per thousand feet.

Variation of the lapse rate is the main reason that *temperatures aloft* are normally measured twice daily. Valuable information can be determined from observations aloft such as freezing level, types of clouds that will form, maximum and minimum surface temperatures for the day, and the amount of air turbulence. The rate of change varies from day to day depending on the amount of heat reaching the Earth, the amount escaping from the Earth, and the type and amount of *advection* (warmer or colder air moving into the area from other regions).

CAUSES OF TEMPERATURE CHANGE

Three important factors are responsible for temperature variation on the earth's surface. They are the Earth's daily rotation about its axis, yearly revolution around the sun, and variations in land mass and water mass heating (Figs. 3-2 through 3-9).

Daily surface heating and cooling results from the Earth's rotation about its axis. As the Earth turns, the side facing the sun is heated and the side away from the sun is cooled. Generally, the lowest temperature occurs near sunrise and the highest temperature is recorded between 1:00 and 4:00 p.m.

The effects of the yearly revolution around the sun, as you learned earlier, are modified by the tilt in the axis of the Earth. Areas under the direct perpendicular rays of the sun receive more heat than those under the oblique rays. One factor affecting this is shown in Fig. 3-10. The lines intersecting the ground represent a group of rays from the sun, which in turn represents a given amount of energy. As these rays strike the ground obliquely, as in winter, the energy is distributed over an area whose width is from *A* to *C*. In summer, when the sun is closer to the overhead position, the same group of rays fall on an area whose width is from *B* to *C*, which is much less than *A* to *C*. Thus, the sun's energy is more concentrated in summer than in winter, which results in greater heating per unit area.

Another reason for the lesser heating in winter is that the oblique rays pass through more atmosphere. This absorbs, reflects, and scatters the sun's energy. Fewer of these oblique rays reach the Earth's surface or lower layers of the atmosphere.

The third factor is variation in land mass and water mass heating. Land areas heat and cool more rapidly than water areas; water tends to have a more uniform temperature throughout the year.

During the night, water retains its warmth while the land mass rapidly loses its heat to space. This difference between land and water heating rates also influences seasonal temperatures. That means oceanic climates are warmer in winter and cooler in summer than land climates at the same latitude.

Another factor in considering temperature and its role in weather forecasting is *heat transfer*. When a solid object or volume of liquid or gas loses more heat energy than it gains, its temperature decreases. When it acquires more heat energy than it loses, its temperature increases. At times, a complete change in the character of the weather occurs over land areas between early morning and midafternoon. This change takes place because of the heating and cooling of the earth and its atmosphere. The four effective methods of heat transfer in the atmosphere are radiation, conduction, convection, and advection.

Radiation is the process that transfers energy through space or a material from one location to another. The sun radiates heat energy to the Earth, called *insolation*. Any reflected energy is called *terrestrial radiation*. The *greenhouse effect* refers to the ability of water vapor, smoke, haze, and particularly clouds to reduce or prevent the cooling of the earth during the night.

Conduction is the transfer of heat by contact. This process is important in meteorology because it causes the air close to the surface of the Earth to heat during the day and cool during the night.

Convection is the heat transferred when air moves upward or downward in the atmosphere, either because of solar heating (*thermal convection*) or air being forced to go over a mountain range (*mechanical convection*).

Advection is the horizontal movement of air (wind) that causes heating or cooling of an area.

Understanding the different factors that change the temperature can help you in understanding weather elements. (For record high and low temperatures, see Tables 3-1 through 3-4.)

PRESSURE

No matter where you stand on this Earth, there's a certain amount of *atmospheric pressure*

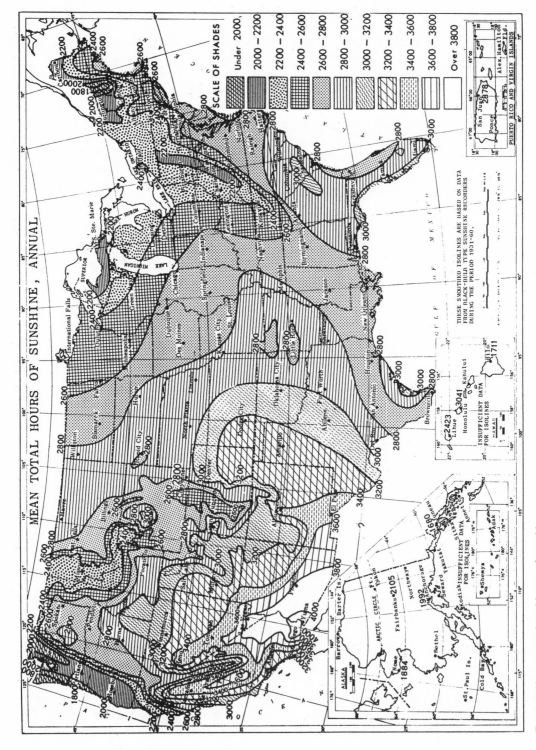

Fig. 3-2. Mean annual total hours of sunshine.

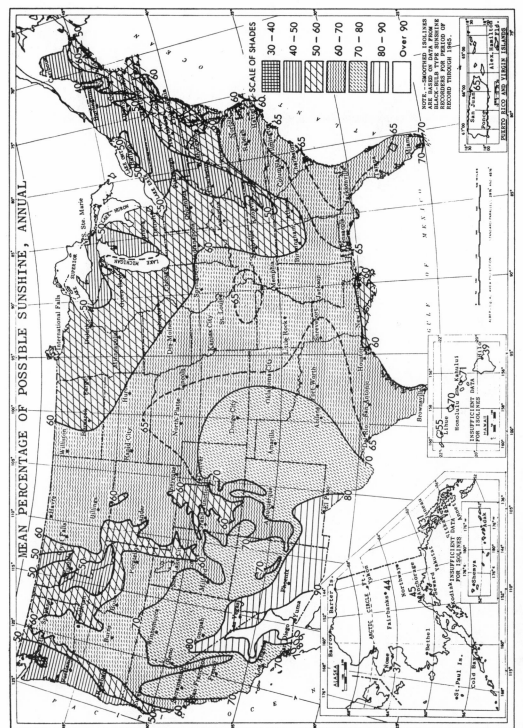

Fig. 3-3. Annual mean percentage of possible sunshine.

41

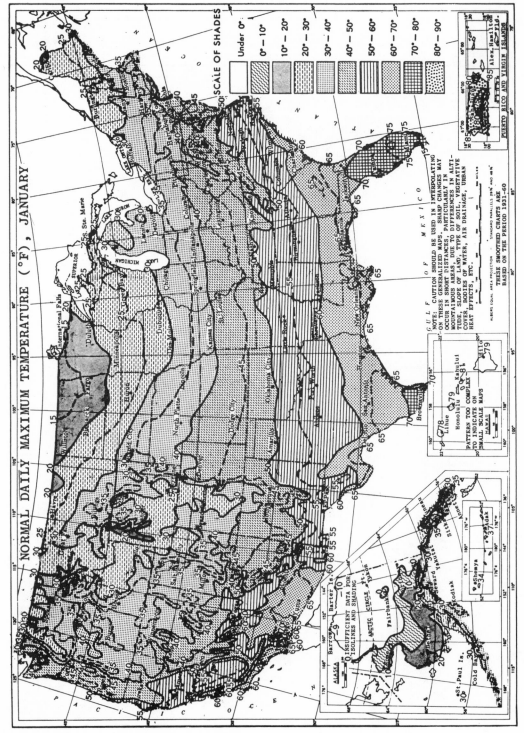

Fig. 3-4. Normal daily maximum temperature (°F) for January.

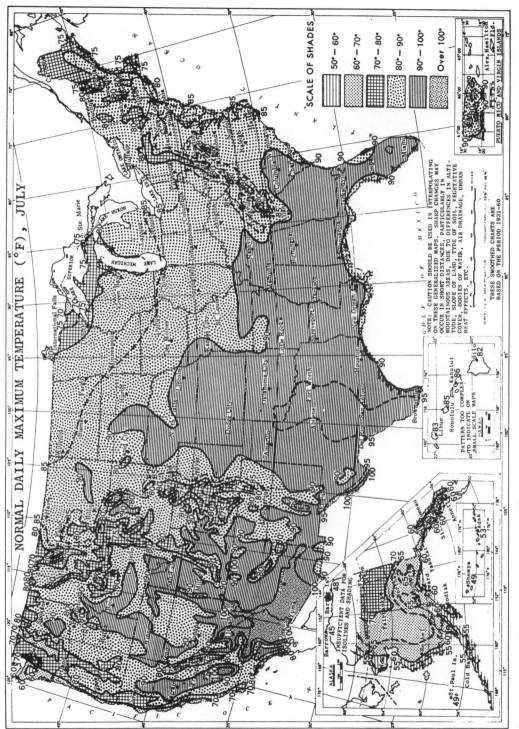

Fig. 3-5. Normal daily maximum temperature (°F) for July.

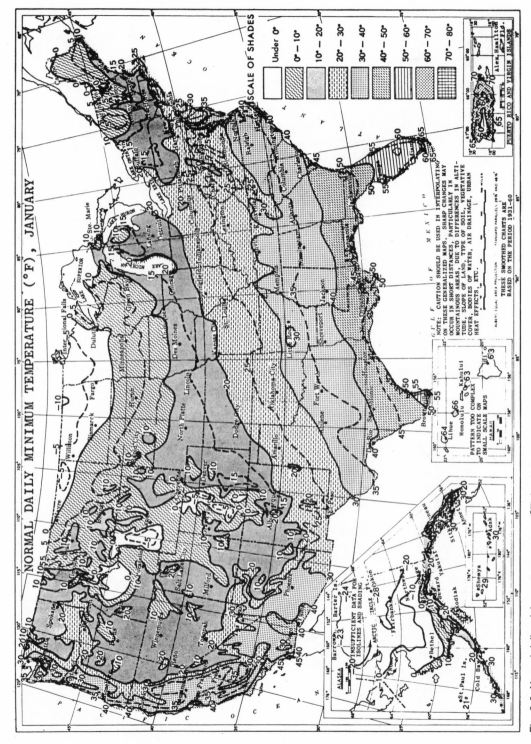

Fig. 3-6. Normal daily minimum temperature (°F) for January.

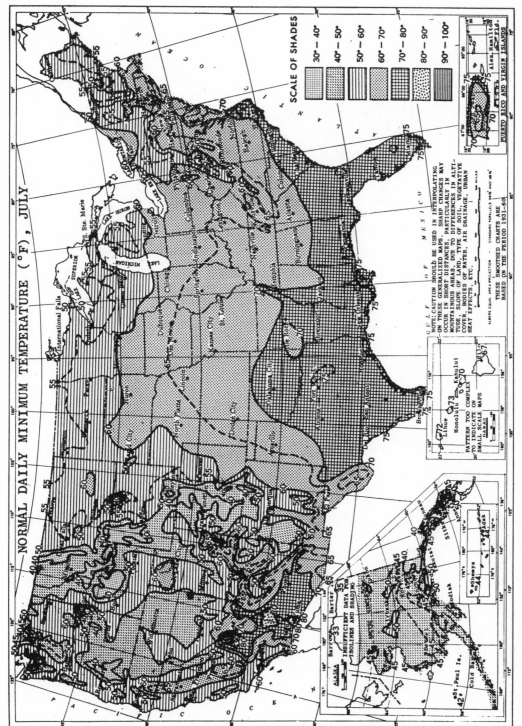

Fig. 3-7. Normal daily minimum temperature (°F) for July.

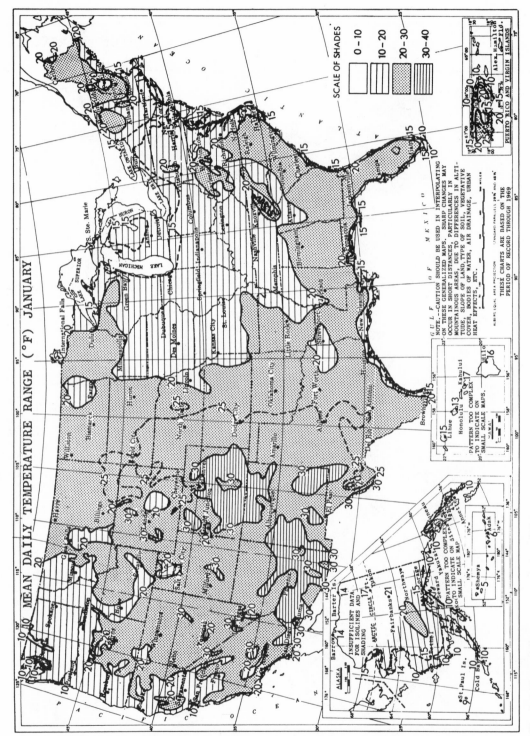

Fig. 3-8. Mean daily temperature range (°F) for January.

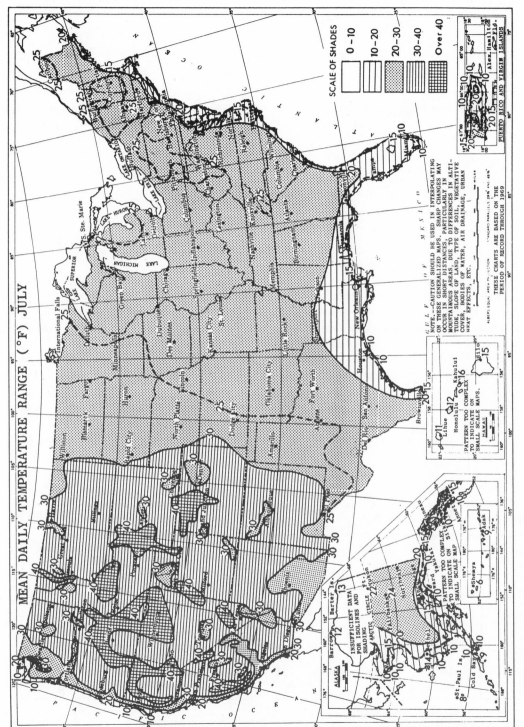

Fig. 3-9. Mean daily temperature range (°F) for July.

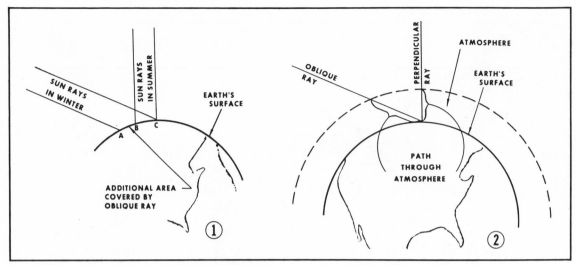

Fig. 3-10. Variations in solar energy received by the earth.

holding you—and everything else—down. Technically, *atmospheric pressure* is the pressure exerted by the atmosphere as a result of gravitational attraction acting on the column of air lying directly above a point. Because of the constant and complex air movements and changes in temperature and moisture content of the air, the weight of the air column over a fixed point is constantly fluctuating. These changes in weight (and therefore pressure) are unfelt by us, but are measurable with pressure-sensitive instruments (Fig. 3-11).

The instruments commonly used in measuring atmospheric pressure are the *mercurial barometer*, the *aneroid barometer* (to be discussed in Chapter 5), and the *barograph* (Chapter 6).

Inches of mercury and *millibars* are two pressure units commonly used. The most common unit is the inch of mercury (abbreviated *Hg*) derived from the height of the mercury column in a mercurial barometer. In meteorology, it is more convenient to express the pressure directly as a force per unit area, the *millibar* (abbreviated *mb*) which is

Table 3-1. Highest Temperatures in the U.S. by Month.

Month	Temp. (°F)	Year	Day	State	Place	Elevation (feet)
Jan.	98	1936	17	Texas	Laredo	421
Feb.*	105	1963	3	Ariz.	Montezuma	735
Mar.*	108	1954	31	Texas	Rio Grande City	168
Apr.	118	1898	25	Calif.	Volcano Springs	−220
May*	124	1896	27	Calif.	Salton	−263
June*†	127	1896	15	Ariz.	Ft. Mohave	555
July	134	1913	10	Calif.	Greenland Ranch	−178
Aug.†	127	1933	12	Calif.	Greenland Ranch	−178
Sept.	126	1950	2	Calif.	Mecca	−175
Oct.*	116	1917	5	Ariz.	Sentinel	685
Nov.*	105	1906	12	Calif.	Croftonville	1,759
Dec.	100	1938	8	Calif.	La Mesa	538

* Two or more occurrences, most recent given.
† Slightly higher temperatures in old records are not used because of lack of information on exposure of instruments.

Table 3-2. Lowest Temperatures in the U.S. by Month.

Month	Temp. (°F)	Year	Day	State	Place	Elevation (feet)
Jan.†	−70	1954	20	Mont.	Rogers Pass	5,470
Feb.	−66	1933	9	Mont.	Riverside R. S.	6,700
Mar.	−50	1906	17	Wyo.	Snake River	6,862
Apr.	−36	1945	5	N. Mex.	Eagle Nest	8,250
May	−15	1964	7	Calif.	White Mountain 2	12,470
June	2	1907	13	Calif.	Tamarack	8,000
July*	10	1911	21	Wyo.	Painter	6,800
Aug.*	5	1910	25	Mont.	Bowen	6,080
Sept.*	−9	1926	24	Mont.	Riverside R. S.	6,700
Oct.	−33	1917	29	Wyo.	Soda Butte	6,600
Nov.	−53	1959	16	Mont.	Lincoln 14 NE	5,130
Dec.*	−59	1924	19	Mont.	Riverside R. S.	6,700

† −80°F, Jan. 23, 1971 at Prospect Creek Camp, Alaska, elevation 1,100 ft.
* Two or more occurrences, most recent given.

atmospheric pressure equal to a force of 1000 dynes per square centimeter. Remember this: Standard atmospheric pressure at sea level is 1013.2 millibars, which is equivalent to 29.92 inches of mercury.

In conversion: The pressure exerted by one inch of mercury is equivalent to approximately 34 millibars, and the pressure exerted by one millibar is equivalent to approximately 0.03 inch of mercury.

To eliminate pressure variations caused by stations being at different altitudes, the mean sea level pressure is plotted in millibars at each reporting station on a surface weather map (Fig. 3-11). Lines (isobars) are drawn connecting equal values of reported mean sea level pressure. Standard procedure on maps of North America is to draw isobars for every 4 millibars (Figs. 3-12 and 3-13).

LOWS, HIGHS, AND PRESSURE GRADIENTS

A low pressure system is one in which the barometric pressure decreases toward the center and the windflow around the system is counterclockwise in the Northern Hemisphere. Low pressure systems with severe storm characteristics are called hurricanes, typhoons, tropical storms, tornadoes, or waterspouts to identify the exact nature of the storm.

A high pressure system is one in which the barometric pressure increases toward the center and the windflow around the system is clockwise in the Northern Hemisphere.

The rate of change in pressure in a direction perpendicular to the isobars is called *pressure gradient*. On a weather map, when the isobars are close together, there is a steep change in atmospheric pressure within a short surface distance. This can be a sign of rough weather. Figures 3-14 and 3-15 illustrate what a high and a low pressure system would look like if you could see them. The gradient or slope of the pressure isobars is a good indicator of the strength and form of incoming weather. It plays a special part in analyzing and forecasting wind.

WIND

Pressure gradients initiate the movement of air. As soon as the air builds up speed, the *Coriolis force* (caused by the Earth's rotation) deflects it to the right in the Northern Hemisphere or to the left in the Southern Hemisphere. As the speed increases along the isobars, this force becomes equal and opposite to the pressure gradient force and causes what's called the *gradient wind* above 3000 feet.

Table 3-3. Highest Recorded Temperatures by State.

State	Temp. (°F)	Date	Station	Elevation (feet)
Ala.	112	Sept. 5, 1925	Centerville	345
Alaska	100	June 27, 1915	Fort Yukon	—
Ariz.	127	July 7, 1905*	Parker	345
Ark.	120	Aug. 10, 1936	Ozark	396
Calif.	134	July 10, 1913	Greenland Ranch	−178
Colo.	118	July 11, 1888	Bennett	—
Conn.	105	July 22, 1926	Waterbury	400
Del.	110	July 21, 1930	Millsboro	20
D.C.	106	July 20, 1930*	Washington	112
Fla.	109	June 29, 1931	Monticello	207
Ga.	112	July 24, 1952	Louisville	337
Hawaii	100	Apr. 27, 1931	Pahala	850
Idaho	118	July 28, 1934	Orofino	1,027
Ill.	117	July 14, 1954	E. St. Louis	410
Ind.	116	July 14, 1936	Collegeville	672
Iowa	118	July 20, 1934	Keokuk	614
Kans.	121	July 24, 1936*	Alton (near)	1,651
Ky.	114	July 28, 1930	Greensburg	581
La.	114	Aug. 10, 1936	Plain Dealing	268
Maine	105	July 10, 1911*	North Bridgton	450
Md.	109	July 10, 1936*	Cumberland & Frederick	623; 325
Mass.	106	July 4, 1911*	Lawrence	51
Mich.	112	July 13, 1936	Mio	963
Minn.	114	July 6, 1936*	Moorhead	904
Miss.	115	July 29, 1930	Holly Springs	600
Mo.	118	July 14, 1954*	Warsaw & Union	687; 560
Mont.	117	July 5, 1937	Medicine Lake	1,950
Nebr.	118	July 24, 1936*	Minden	2,169
Nev.	122	June 23, 1954*	Overton	1,240
N.H.	106	July 4, 1911	Nashua	125
N.J.	110	July 10, 1936	Runyon	18
N. Mex.	116	July 14, 1934*	Orogrande	4,171
N.Y.	108	July 22, 1926	Troy	35
N.C.	109	Sept. 7, 1954*	Weldon	81
N. Dak.	121	July 6, 1936	Steele	1,857
Ohio	113	July 21, 1934*	Gallipolis (near)	673
Okla.	120	July 26, 1943*	Tishomingo	670
Oreg.	119	Aug. 10, 1898*	Pendleton	1,074
Pa.	111	July 10, 1936*	Phoenixville	100
R.I.	102	July 30, 1949	Greenville	420
S.C.	111	June 28, 1954*	Camden	170
S. Dak.	120	July 5, 1936	Gannvalley	1,750
Tenn.	113	Aug. 9, 1930*	Perryville	377
Tex.	120	Aug. 12, 1936	Seymour	1,291
Utah	116	June 28, 1892	Saint George	2,880
Vt.	105	July 4, 1911	Vernon	310
Va.	110	July 15, 1954	Balcony Falls	725
Wash.	118	Aug. 5, 1961*	Ice Harbor Dam	475
W. Va.	112	July 10, 1936*	Martinsburg	435
Wis.	114	July 13, 1936	Wisconsin Dells	900
Wyo.	114	July 12, 1900	Basin	3,500
P.R.	103	Aug. 22, 1906	San Lorenzo	203

* Also on earlier dates at the same or other places in the state.

Table 3-4. Lowest Recorded Temperatures by State.

State	Temp. (°F)	Date	Station	Elevation (feet)
Ala.	−24	Jan. 31, 1966	Russellville	880
Alaska	−80	Jan. 23, 1971	Prospect Creek Camp	1,100
Ariz.	−40	Jan. 7, 1971	Hawley Lake	8,180
Ark.	−29	Feb. 13, 1905	Pond	1,250
Calif.	−45	Jan. 20, 1937	Boca	5,532
Colo.	−60	Feb. 1, 1951	Taylor Park	9,206
Conn.	−32	Feb. 16, 1943	Falls Village	585
Del.	−17	Jan. 17, 1893	Millsboro	20
D.C.	−15	Feb. 11, 1899	Washington	112
Fla.	− 2	Feb. 13, 1899	Tallahassee	193
Ga.	−17	Jan. 27, 1940	CCC Camp F-16	—
Hawaii	14	Jan. 2, 1961	Haleakala, Maui Island	9,750
Idaho	−60	Jan. 18, 1943	Island Park Dam	6,285
Ill.	−35	Jan. 22, 1930	Mount Carroll	817
Ind.	−35	Feb. 2, 1951	Greensburg	954
Iowa	−47	Jan. 12, 1912	Washta	1,157
Kans.	−40	Feb. 13, 1905	Lebanon	1,812
Ky.	−34	Jan. 28, 1963	Cynthiana	684
La.	−16	Feb. 13, 1899	Minden	194
Maine	−48	Jan. 19, 1925	Van Buren	510
Md.	−40	Jan. 13, 1912	Oakland	2,461
Mass.	−34	Jan. 18, 1957	Birch Hill Dam	840
Mich.	−51	Feb. 9, 1934	Vanderbilt	785
Minn.	−59	Feb. 16, 1903*	Pokegama Dam	1,280
Miss.	−19	Jan. 30, 1966	Corinth 4 SW	420
Mo.	−40	Feb. 13, 1905	Warsaw	700
Mont.	−70	Jan. 20, 1954	Rogers Pass	5,470
Nebr.	−47	Feb. 12, 1899	Camp Clarke	3,700
Nev.	−50	Jan. 8, 1937	San Jacinto	5,200
N.H.	−46	Jan. 28, 1925	Pittsburg	1,575
N.J.	−34	Jan. 5, 1904	River Vale	70
N. Mex.	−50	Feb. 1, 1951	Gavilan	7,350
N.Y.	−52	Feb. 9, 1934	Stillwater Reservoir	1,670
N.C.	−29	Jan. 30, 1966	Mt. Mitchell	6,525
N. Dak.	−60	Feb. 15, 1936	Parshall	1,929
Ohio	−39	Feb. 10, 1899	Milligan	800
Okla.	−27	Jan. 18, 1930*	Watts	958
Oreg.	−54	Feb. 10, 1933*	Seneca	4,700
Pa.	−42	Jan. 5, 1904	Smethport	—
R.I.	−23	Jan. 11, 1942	Kingston	100
S.C.	−13	Jan. 26, 1940	Longcreek (near)	1,631
S. Dak.	−58	Feb. 17, 1936	McIntosh	2,277
Tenn.	−32	Dec. 30, 1917	Mountain City	2,471
Tex.	−23	Feb. 8, 1933*	Seminole	3,275
Utah	−50	Jan. 5, 1913*	Strawberry Tunnel	7,650
Vt.	−50	Dec. 30, 1933	Bloomfield	915
Va.	−29	Feb. 10, 1899	Monterey	—
Wash.	−48	Dec. 30, 1968	Mazama & Winthrop	2,120; 1,755
W. Va.	−37	Dec. 30, 1917	Lewisburg	2,200
Wis.	−54	Jan. 24, 1922	Danbury	908
Wyo.	−63	Feb. 9, 1933	Moran	6,700
P.R.	40	Mar. 9, 1911	Aibonito	2,059

* Also on earlier dates at the same or other places in the state.

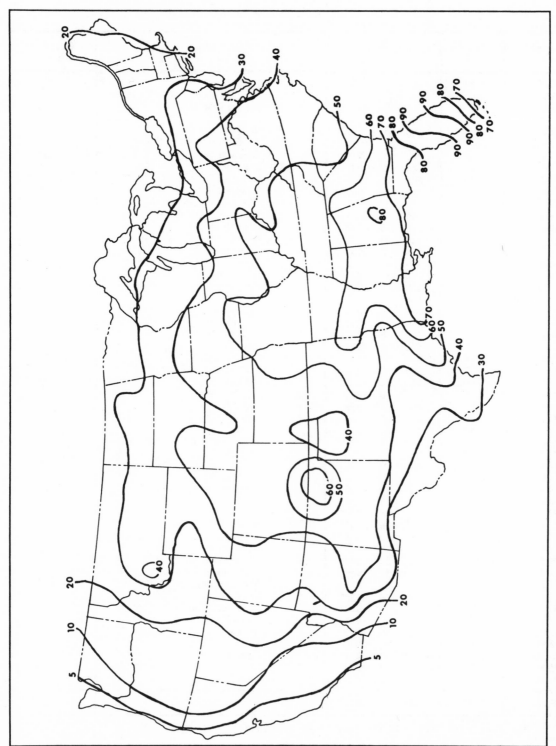

Fig. 3-11. Average number of thunderstorms each year over the conterminous United States.

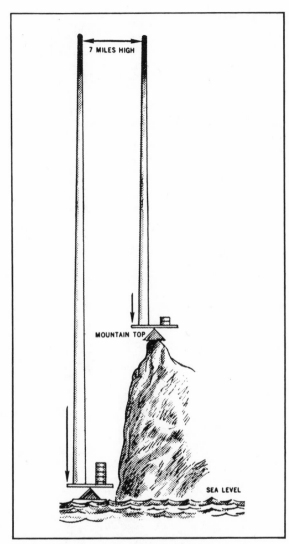

Fig. 3-12. Air pressure varies with elevation.

Friction reduces the surface wind speed to about 40 percent of the velocity of the gradient wind and causes the surface wind to flow across the isobars instead of parallel to them. The amount of friction—and thus the speed of the surface wind—depends on the nature of the surface. The friction is least over water and greatest over mountainous terrain. The average surface wind will flow *across* the isobars *toward* lower pressure at about a 30° angle. The surface friction gradually decreases with altitude until the gradient level is reached.

Surface winds flow clockwise around and away from a center of high pressure and counter-clockwise around and toward a center of low pressure in the Northern Hemisphere.

Other winds and circulation factors come into play to keep meteorologists—professional and amateur—guessing. During daytime, coastal land generally becomes warmer than the adjacent water and a lower density will exist in the surface layer of air over land than in the surface layer over the water (Fig. 3-15). This slight difference in pressure over the land and water surfaces establishes a flow of wind landward (a sea breeze) during the day. The force of the sea breezes depends on the amount of insolation and terrestrial radiation. Sea breezes are most pronounced on clear days, in the summer and in low latitudes. Land breezes (from land to sea) occur at night due to the rapid nocturnal cooling of the land surface.

On warm days, winds tend to flow up slopes during the day and down slopes during the night. This is because air in contact with mountain slopes is warmer than the free atmosphere at the same level during the day and colder during the night. Since cold air tends to sink and warm air tends to rise, a system of winds develops and flows up the mountain side during the day and down during the night. The daytime movement is a valley breeze and the nighttime motion is a mountain breeze (Figs. 3-16 through 3-18).

For prevailing world wind patterns, see Fig. 3-19.

THE JETSTREAM

The jetstream is a special circulation, very important to understanding weather. The term *jetstream* is defined as relatively strong winds concentrated within a narrow stream in the atmosphere (Fig. 3-20). While this term may be applied to any such stream regardless of direction, it is commonly interpreted to mean a band or belt of winds with a strong westerly component that meanders around the globe. By saying that it has a strong westerly component, it is meant that it flows primarily from the west or from adjacent directions such as north-

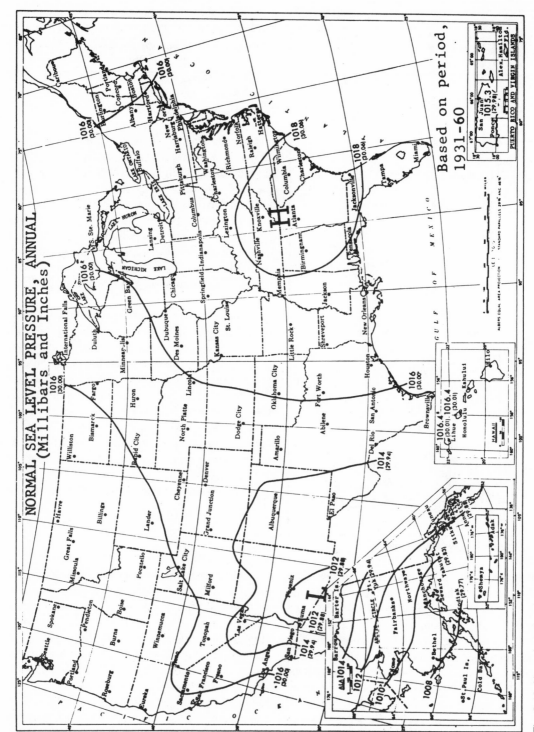

Fig. 3-13. Normal sea level pressure, annually.

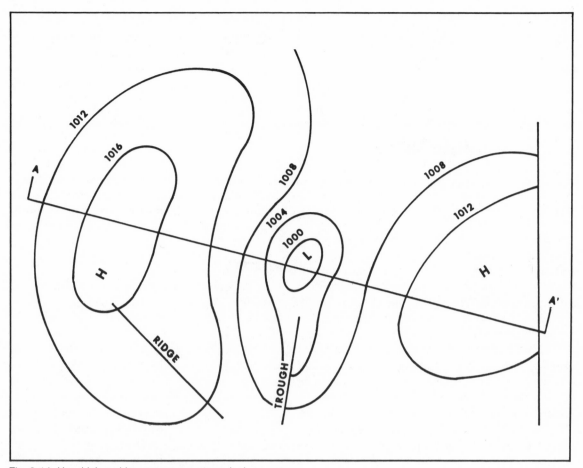

Fig. 3-14. How high and low pressure systems look on a map.

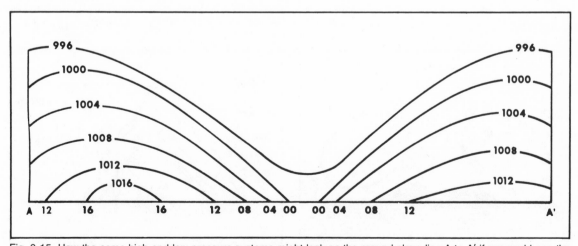

Fig. 3-15. How the same high and low pressure systems might look on the ground along line A to A' if you could see the pressure gradients.

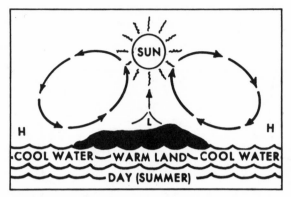

Fig. 3-16. Land-sea effect during the day.

west or southwest. By saying that it meanders, it is meant that it is not found at the same latitude or elevation all around the earth at the same time, but it has a wavelike trajectory. It may range from 25 to 100 miles in width and up to a mile or two in depth. Sometimes the jetstream is a continuous band, but more often it is broken or split at several points.

Jetstreams are found in both the Northern Hemisphere and the Southern Hemisphere, but much more is known about the predominant one in the Northern Hemisphere. This is the one normally referred to when only the term "jetstream" is used. It is located in the high tropopause along the boundary of the polar front zone where there is extreme horizontal temperature contrast. Normally, there is a break in the tropopause where the jetstream exists, or it may be said that it exists where the tropopause has its greatest slope (Fig. 3-21).

The winds in the jetstream occasionally exceed 250 knots (288 miles an hour). Most of the time the winds range from about 100 to 150 knots (115 to 173 miles an hour). However, a band of winds is classed as a jetstream only when the winds in the band have a speed of 50 knots (58 miles an hour) or more. The jetstream is stronger in winter than in summer.

The jetstream is closely associated with migratory low pressure systems and the polar front. It is very important in forecasting weather relative to the development and movement of fronts and low pressure systems. It is also important to aviators as an aid or hindrance to flights above 40,000 feet.

MOISTURE

Moisture is a major element of what we call "weather." Moisture causes rain and snow, high and low humidity. The atmosphere's level and form of moisture can cause comfort or discomfort.

Atmospheric moisture comes in three states: vapor, liquid, and solid. Moisture vapor is called fog, liquid is rain, and solid is ice, snow, sleet, or hail, depending on its exact state. Like everything else in weather, moisture must live with certain natural rules, which include evaporation, condensation, sublimation, and heat exchange.

Evaporation is a change in state from a liquid to a gas. A pan of water set out on a warm day will evaporate—change from a liquid to a gas—into the air (Fig. 3-22). It's the same with water in a lake or

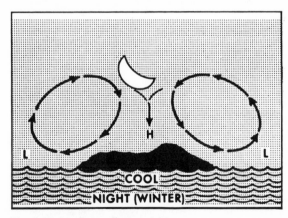

Fig. 3-17. Land-sea effect at night.

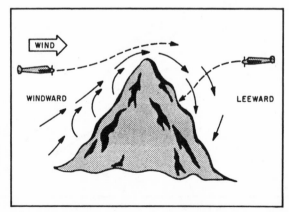

Fig. 3-18. Effect of windflow over mountains.

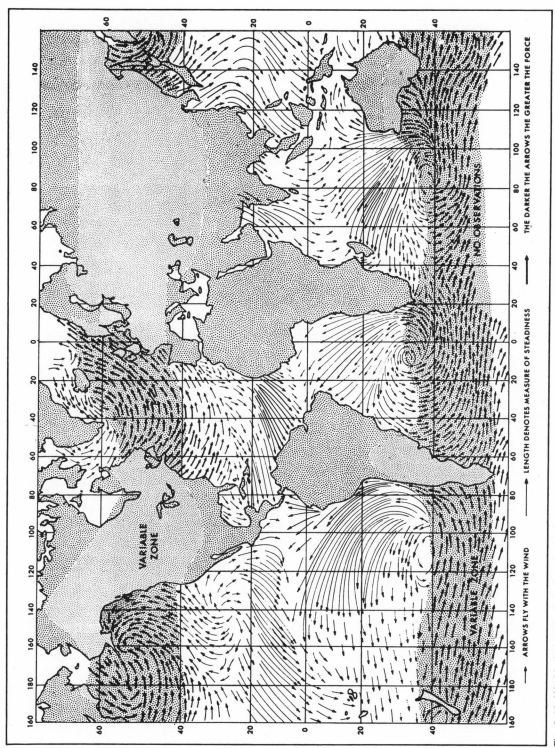

Fig. 3-19. World wind patterns during the winter.

57

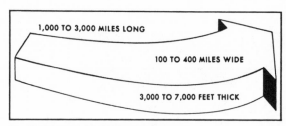

1,000 TO 3,000 MILES LONG

100 TO 400 MILES WIDE

3,000 TO 7,000 FEET THICK

Fig. 3-20. Size of a jetstream segment.

an ocean. As the temperature of the liquid increases, more of the molecules will obtain escape velocity and the rate of evaporation will increase—depending on the air pressure and temperature as well as the temperature of the liquid.

Condensation is the reverse of evaporation; it is the change of state from a gas to a liquid. Condensation occurs when the air holding the moisture reaches a saturation point and can no longer hold the moisture in a gaseous form. The moisture is turned into water droplets—or, if the air is cold enough, ice. This is what makes rain and snow.

Sublimation is the direct change from solid to vapor and vice versa without passing through the intermediate liquid state. This occurs in clouds at below freezing temperatures as ice and snow form directly from water vapor in the air without first becoming a liquid.

Heat exchange is the amount of heat needed to change moisture from a gas to a liquid to a solid and back again (Fig. 3-23).

Those are the ground rules for moisture. Now let's look at the specifics of how they react in our atmosphere to give us sunny days and snowstorms.

Saturation is a condition that exists when air at a given temperature and pressure is holding the maximum possible water vapor content. In the atmosphere, saturation occurs by evaporation into the air from a free water surface (ocean, lake, river, rain drops) until vapor pressure is equal between water and air. *Dew point* is that temperature to which air must be cooled to become saturated. For example, if the air temperature were 60°F and the dew point 50°F, the air would be saturated if cooled to 50°F. If the air were further cooled to 49°F, it could no longer hold all the water vapor present and

some of it would be virtually "squeezed out" in the form of liquid water—condensation (Fig. 3-24).

The moisture content of the air is measured by expressing the humidity as the amount of water vapor (in grams) contained in one kilogram of natural air (Table 3-5). This is called *specific humidity* (Fig. 3-25). But most of the atmosphere is not saturated. It contains less than the maximum possible quantity of water vapor (Fig. 3-26). For weather analysis it's best to express how *near* the air is to being saturated. This is the *relative humidity*—the ratio of the amount of water vapor in the air and the amount of water vapor that the air would contain when saturated at the same temperature (Figs. 3-27, 3-28). It is expressed as a percentage, with saturated air having 100 percent relative humidity. It's a handy figure to know. Figure 3-29 shows how the dew point and the relative humidity are related.

PRECIPITATION

Precipitation is the general term for all forms of falling moisture—rain, snow, hail, ice pellets, etc. (Fig. 3-30). The chain of events that lead up to precipitation include saturation of the air, condensation of moisture into clouds, and heat exchange of cloud vapor into liquid or solids.

Precipitation then occurs in the form of liquid or frozen moisture. For average precipitation, see Fig. 3-31.

Rain is precipitation that reaches the earth's surface as relatively large droplets. Rain can be classified as light, moderate, or heavy, based on the rate of fall or the effect it has on surfaces or visibility. *Drizzle* is precipitation from stratiform clouds in smaller droplets, a sign of little or no turbulence in the air. Drizzle, too, is classified as light, moderate, or heavy.

Freezing rain or *drizzle* is precipitation in the form of supercooled liquid raindrops, a portion of which freezes and forms a smooth coating of ice upon striking the surface (Fig. 3-32). Frozen precipitation comes in three styles: *ice pellets* (Fig. 3-33) or *sleet* (frozen raindrops formed by rain or drizzle falling through a below-freezing layer of air), *hail* (Fig. 3-34) (lumps or balls of ice circulated

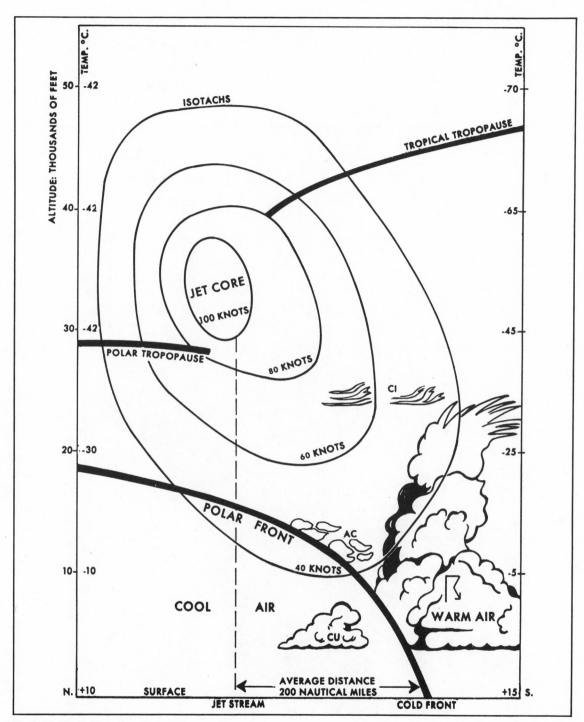

Fig. 3-21. Vertical cross section of a jetstream.

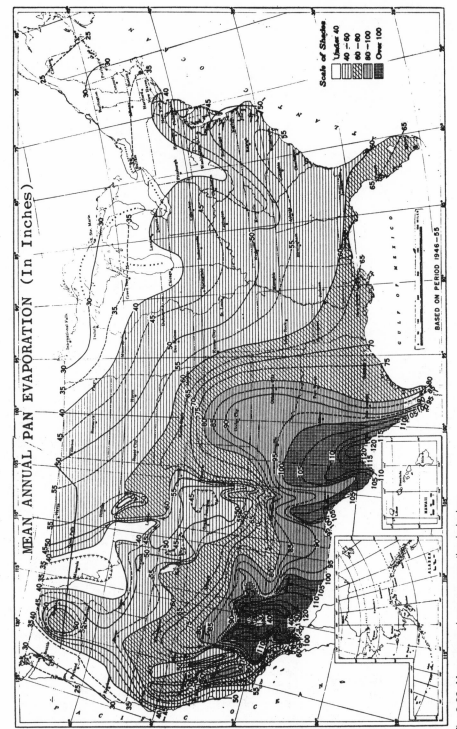

Fig. 3-22. Mean annual pan evaporation in inches.

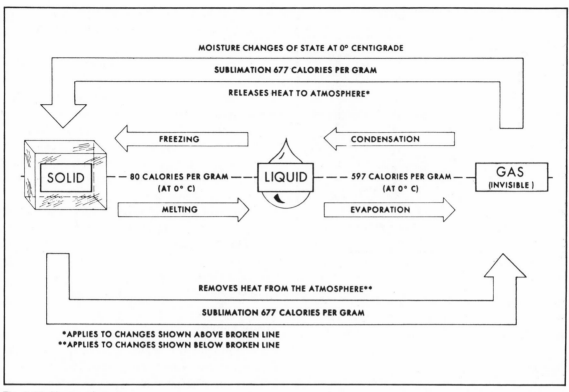

Fig. 3-23. Heat exchanges.

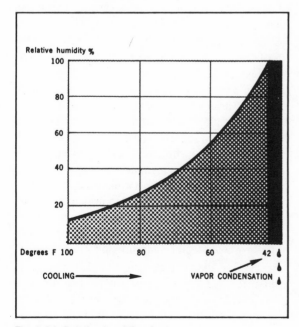

Fig. 3-24. Relative humidity chart.

Table 3-5. Relative Humidity/Temperature Table.

Temp. °F	Relative humidity %		
	20	30	40
	Equilibrium moisture content (percent)		
40	4.6	6.3	7.9
45	4.6	6.3	7.9
50	4.6	6.3	7.8
55	4.6	6.2	7.8
60	4.6	6.2	7.7
65	4.5	6.1	7.7
70	4.5	6.1	7.6
75	4.5	6.0	7.5
80	4.4	5.9	7.4
85	4.4	5.9	7.3
90	4.3	5.8	7.2
95	4.3	5.7	7.1
100	4.2	5.6	7.0
103	4.1	5.5	6.9
110	4.0	5.4	6.8
115	3.9	5.3	6.7
120	3.8	5.2	6.5

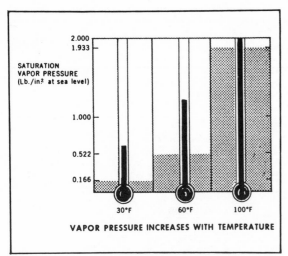

Fig. 3-25. Capacity of air to hold water is relative to temperature.

through freezing clouds adding layers of ice to sizes up to 2 pounds), and *snow* (ice crystals formed by sublimation in below-freezing clouds (Fig. 3-35).

CLOUDS

Cloud formations are the direct result of saturation-producing processes that take place in the atmosphere. Identifying and interpreting clouds can help you in understanding and forecasting weather.

Clouds are visible condensed moisture consisting of droplets of water or crystals of ice. They are easily supported and transported by air movements as slow as one-tenth of a mile per hour.

The international cloud classification (Table 3-36) is designed primarily to provide a standardized cloud classification. Within this classification, cloud types are usually divided into four major groups and further classified in terms of their forms and appearance. The four major groups are high clouds, middle clouds, low clouds, and vertical clouds.

Within the high, middle, and low cloud groups there are two main subdivisions. Clouds formed when localized vertical currents carry moist air upward to the condensation level are called *cumuliform*-type clouds, meaning "accumulation" or "heap." These clouds are characterized by their lumpy or billowy appearance. Fliers know that turbulent flying conditions are usually found in and below cumuliform clouds.

Stratiform-type clouds—meaning "spread out"—are formed when complete layers of air are cooled until condensation takes place. Stratiform clouds usually look like white sheets.

In addition to the two main subdivisions mentioned, the word *nimbus*, meaning "raincloud," is added to the names of clouds that normally produce heavy precipitation. For example, a stratiform cloud producing precipitation is referred to as *nimbostratus*, and a heavy, swelling cumulus cloud that has grown into a thunderstorm is referred to as *cumulonimbus*. Clouds that are broken into fragments are identified by adding the prefix *fracto* to the classification name. For example, fragmentary cumulus is referred to as *fractocumulus*.

CLOUD TYPES

The high cloud group consists of *cirrus, cirocumulus,* and *cirrostratus* clouds. The mean base level of these three cloud types is 20,000 feet or higher above terrain. Cirrus clouds give indications of approaching changes in weather. Cirriform clouds are composed of ice crystals. These clouds are usually thin and the outline of the sun or moon may be seen through them, producing a halo effect.

The middle cloud group consists of *altocumulus* and *altostratus* clouds. The altocumulus has many variations in appearance and in formation, whereas the altostratus varies mostly in thickness from very thin to several thousand feet. Bases of the middle clouds range from 6500 to 20,000 feet above the terrain. These clouds may be composed of ice crystals or water droplets (which may be supercooled), and may contain icing conditions hazardous to aircraft. Altocumulus rarely produces precipitation, but altostratus usually indicates the proximity of unfavorable weather and precipitation.

The low cloud group consists of *stratus, stratocumulus*, and *nimbostratus* clouds. The bases of these clouds range from near the surface to about 6500 feet above the terrain. The heights of the cloud bases may change rapidly. If low clouds form below

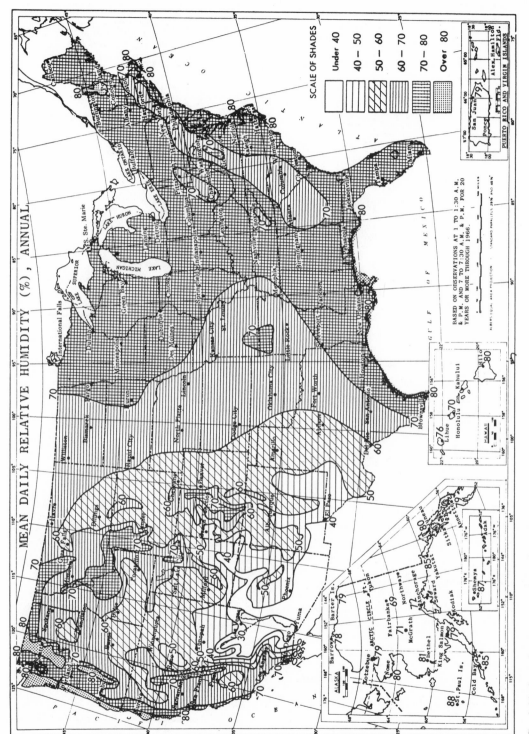

Fig. 3-26. Annual mean daily relative humidity.

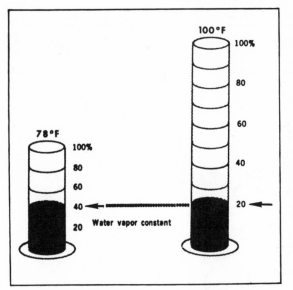

Fig. 3-27. A temperature drop gives rise in relative humidity.

Fig. 3-28. Estimating relative humidity.

$$\frac{\text{ACTUAL MIXING RATIO}}{\text{SATURATION MIXING RATIO}} \times 100 = \begin{array}{c}\text{RELATIVE} \\ \text{HUMIDITY} \\ \text{(\%)}\end{array}$$

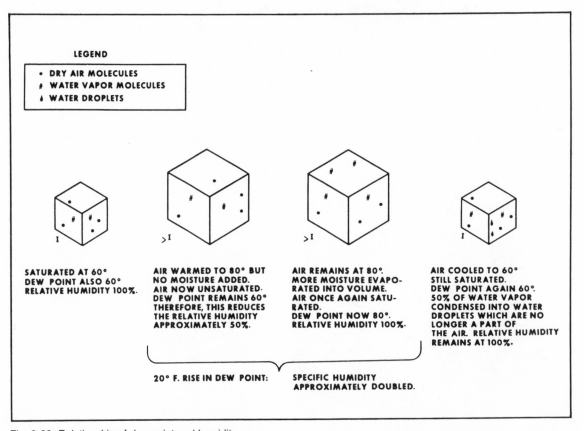

Fig. 3-29. Relationship of dew point and humidity.

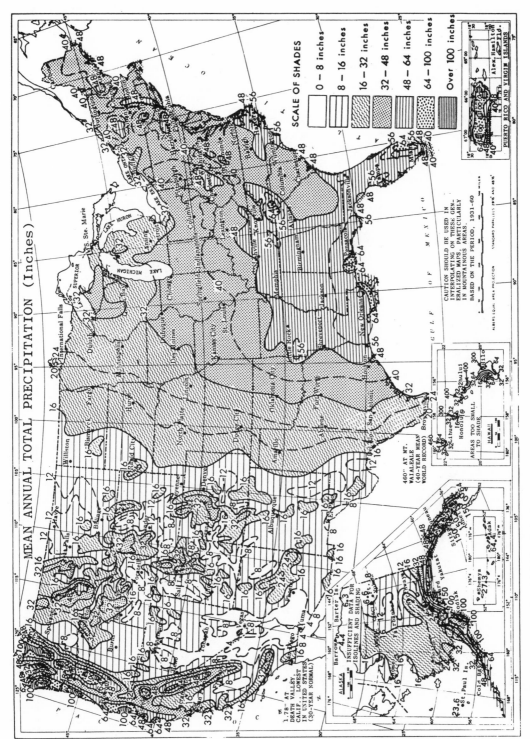

Fig. 3-30. Mean annual total precipitation in inches.

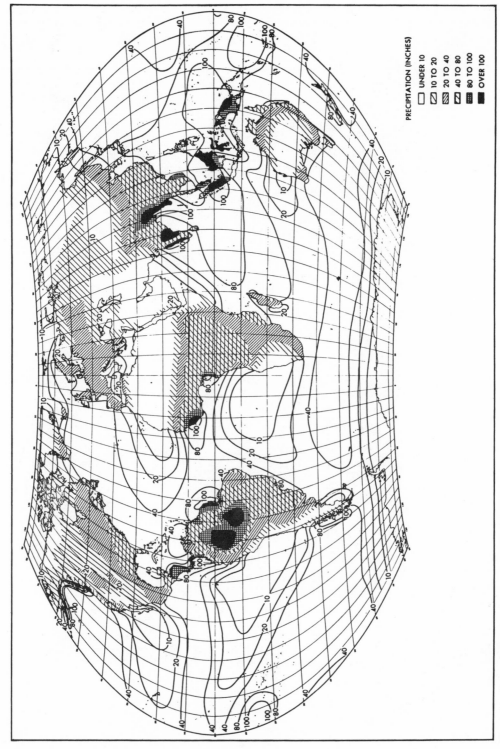

Fig. 3-31. General pattern of annual world precipitation.

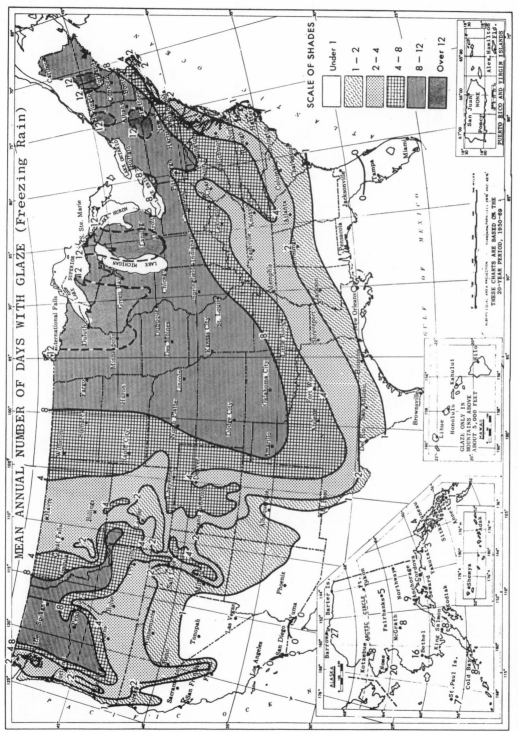

Fig. 3-32. Mean annual number of days with glaze, or freezing rain.

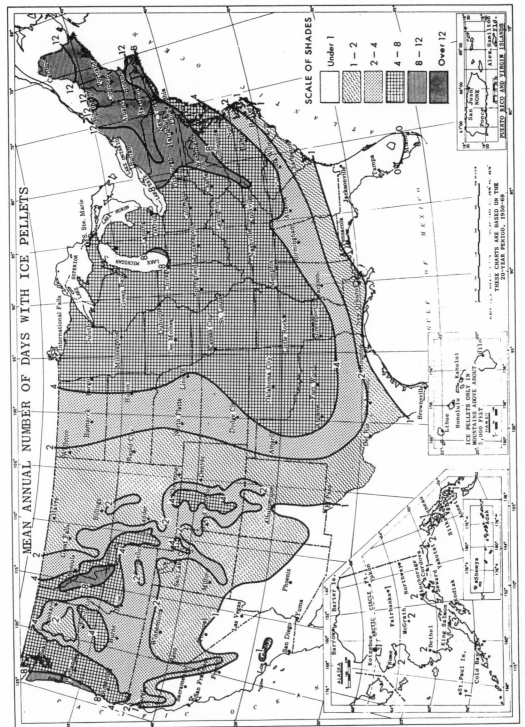

Fig. 3-33. Mean annual number of days with ice pellets.

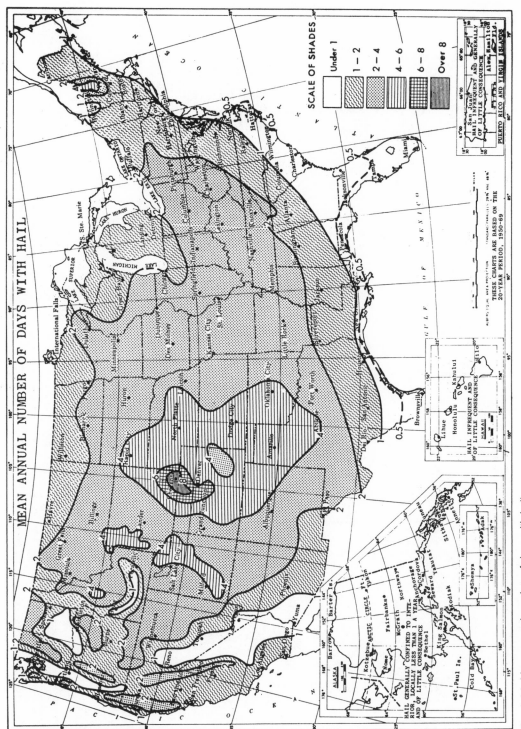

Fig. 3-34. Mean annual number of days with hail.

69

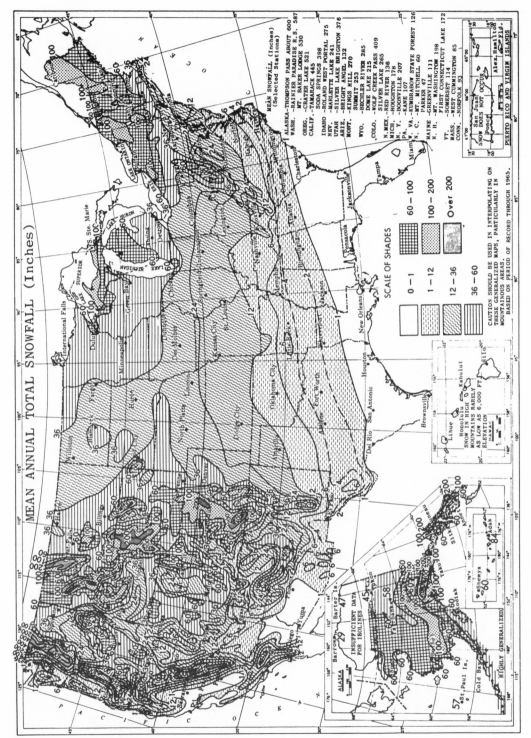

Fig. 3-35. Mean annual total snowfall (inches).

Table 3-6. International Cloud Classification, Abbreviations, and Weather Map Symbols.

Base Altitude	Cloud Type	Abbreviation	Symbol
Bases of high clouds usually above 20,000 feet.	Cirrus	Ci	
	Cirrocumulus	Cc	
	Cirrostratus	Cs	
—— 20,000 feet ——			
Bases of middle clouds range from 6,500 feet to 20,000 feet	Altocumulus	Ac	
	Altostratus	As	
—— 6,500 feet ——			
Bases of low clouds range from surface to 6,500 feet.	*Cumulus	Cu	
	*Cumulonimbus	Cb	
	Nimbostratus	Ns	
	Stratocumulus	Sc	
	Stratus	St	
—— Surface ——			

*Cumulus and Cumulonimbus are clouds with vertical development. Their base is usually below 6,500 feet but may be slightly higher. The tops of the Cumulonimbus sometimes exceed 60,000 feet.

50 feet, they are reclassified as fog and may completely blanket landmarks. Low clouds have the same composition as middle clouds.

Clouds with vertical development include the *cumulus* and *cumulonimbus* clouds. These clouds generally have their bases below 6500 feet above the terrain and tops sometimes exceed 60,000 feet. Clouds with vertical development are caused by some type of lifting action, such as convective currents, convergence, orographic lift, or frontal lift.

The cumulonimbus cloud is what is commonly called the *thunderstorm* cloud. Thunderstorms are almost always accompanied by strong gusts of wind and severe turbulence, heavy rain showers, lightning, and other surprises. Frontal thunderstorms are caused by the lifting of warm, moist, conditionally unstable air over a frontal surface. Thunderstorms may also occur many miles ahead of a rapidly moving cold front. Air mass thunderstorms are contained within air masses and can occur after

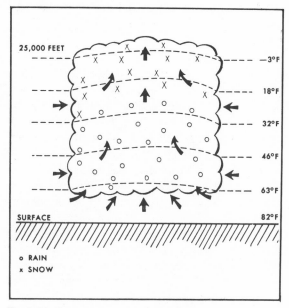

Fig. 3-36. Cumulus stage of a thunderstorm cell.

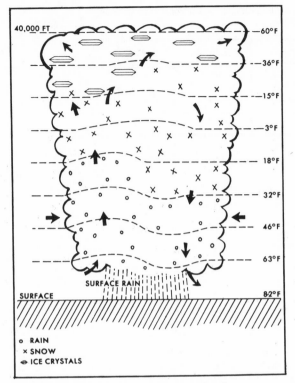

Fig. 3-37. Mature stage of a thunderstorm cell.

a front has moved through. Figures 3-36 through 3-38 show the stages of thunderstorm development.

TORNADOS

A *tornado* is an exceedingly violent whirling storm with a small diameter, usually a quarter of a mile or less. The length of the track of a tornado on the ground may be from a few hundred feet to 300 miles. The average is less than 25 miles. When not touching the ground it is called a *funnel cloud*. Wind velocity within a tornado usually ranges between 150 and 300 miles an hour while it tracks along the ground at a comparatively slow 25 to 40 miles an hour.

Most tornadoes in the United States occur in the late spring and early summer, and are usually associated with thunderstorm activity and heavy rains (Fig. 3-39). The majority of tornados appear about 75 to 180 miles ahead of a cold front along the pre-frontal squall line (Fig. 3-40).

WATERSPOUTS

Waterspouts are tornados that form over ocean areas. One type is a true waterspout in which the vortex forms at the cloud and extends to the surface, usually in advance of a squall line. The pseudo-waterspout originates just above the water surface and builds upwards, much like dust devils found in deserts.

THE HYDROLOGIC CYCLE

Water at the Earth's surface evaporates by absorbing heat energy. The water vapor is transported horizontally and vertically by atmospheric currents until it releases its energy to the atmosphere through condensation that forms clouds and fog. The water cycle is completed when precipitation returns the water to the surface. This evaporation/condensation cycle is a part of the thermodynamics involved in the series of water phenomena called the *hydrologic cycle.*

Man's desire to understand this cycle of life is not new. The Biblical Book of Job says "He (God)

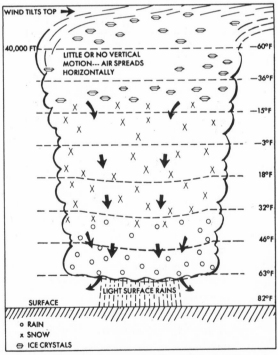

Fig. 3-38. Dissipating stage of a thunderstorm cell.

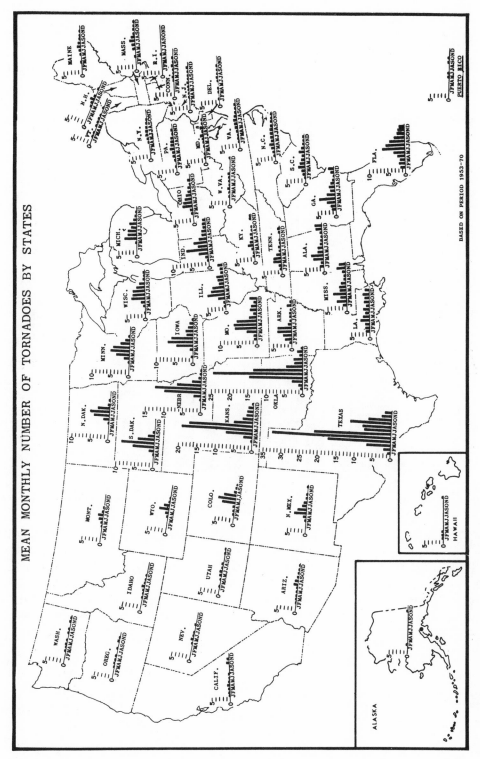

Fig. 3-39. Mean monthly number of tornadoes by state.

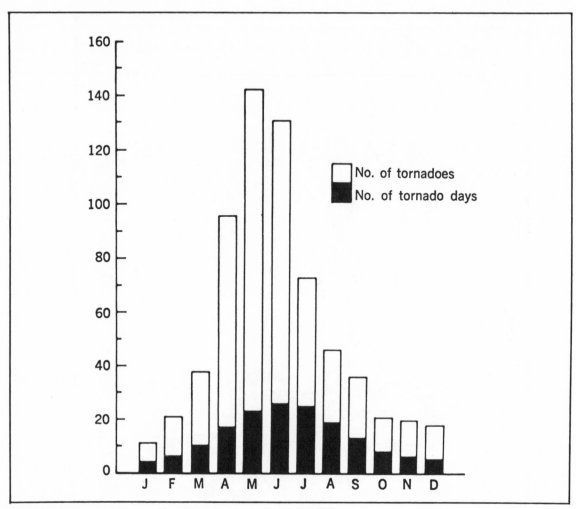

Fig. 3-40. Average monthly number of tornado days in the United States.

draws up the drops of water, which distill as rain to the streams; the clouds pour down their moisture and abundant showers fall on mankind. Who can understand how he spreads out the clouds, how he thunders from his pavilion?" (Job 36:27-29/New International Version).

Our understanding of the miracles of the atmosphere is still small, but, with knowledge from observation, we can become imperfect weather prophets.

(For climatic records from around the world see Fig. 3-41 and the accompanying Table 3-6.)

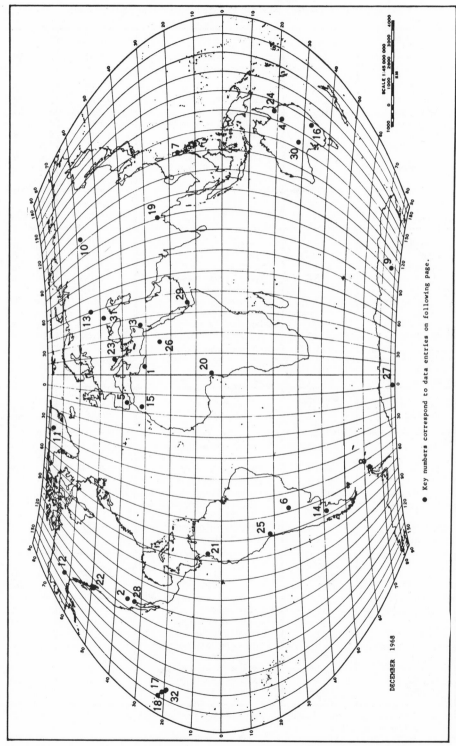

Fig. 3-41. Worldwide extremes of temperature and precipitation recorded by continental area.

● Key numbers correspond to data entries on following page.

DECEMBER 1968

Key No.	Area	Highest °F	Place	Elevation Feet	Date
1	Africa	136	Azizia, Libya	380	Sept. 13, 1922
2	North America	134	Death Valley, Calif.	−178	July 10, 1913
3	Asia	129	Tirat Tsvi, Israel	−722	June 21, 1942
4	Australia	128	Cloncurry, Queensland	622	Jan. 16, 1889
5	Europe	122	Seville, Spain	26	Aug. 4, 1881
6	South America	120	Rivadavia, Argentina	676	Dec. 11, 1905
7	Oceania	108	Tuguegarao, Philippines	72	Apr. 29, 1912
8	Antarctica	58	Esperanza, Palmer Pen.	26	Oct. 20, 1956

Key No.	Area	Lowest °F	Place	Elevation Feet	Date
9	Antarctica	−127	Vostok	11,220	Aug. 24, 1960
10	Asia	−90	Oymykon, U.S.S.R.	2,625	Feb. 6, 1933
11	Greenland	−87	Northice	7,690	Jan. 9, 1954
12	North America	−81	Snag, Yukon, Canada	1,925	Feb. 3, 1947
13	Europe	−67	Ust'Shchugor, USSR	279	January +
14	South America	−27	Sarmiento, Argentina	879	June 1, 1907
15	Africa	−11	Ifrane, Morocco	5,364	Feb. 11, 1935
16	Australia	−8	Charlotte Pass, N.S.W.	---	July 22, 1947*
17	Oceania	14	Haleakala Summit, Maui	9,750	Jan. 2, 1961

+ exact date unknown; lowest in 15-year period
* an earlier date
--- elevation unknown

Key No.	Area	Greatest Amount Inches	Place	Elevation Feet	Years of Record
18	Oceania	460.0	Mt. Waialeale, Kauai, Hawaii	5,075	32
19	Asia	450.0	Cherrapunji, India	4,309	74
20	Africa	404.6	Debundscha, Cameroon	30	32
21	South America	353.9	Quibdo, Colombia	240	10-16
22	North America	262.1	Henderson Lake, B. C., Canada	12	14
23	Europe	182.8	Crkvica, Yugoslavia	3,337	22
24	Australia	179.3	Tully, Queensland	---	31

Key No.	Area	Least Amount Inches	Place	Elevation Feet	Years of Record
25	South America	0.03	Arica, Chile	95	59
26	Africa	<0.1	Wadi Halfa, Sudan	410	39
27	Antarctica	* 0.8	South Pole Station	9,186	10
28	North America	1.2	Batagues, Mexico	16	14
29	Asia	1.8	Aden, Arabia	22	50
30	Australia	4.05	Mulka, South Australia	---	34
31	Europe	6.4	Astrakhan, USSR	45	25
32	Oceania	8.93	Puako, Hawaii	5	13

* The value given is the average amount of solid snow accumulating in one year as indicated by snow markers. The liquid content of the snow is undetermined.

Chapter 4

The Weather Eye

Do-it-yourself weather forecasts have been made throughout the ages. As early as 400 B.C. the Greek philosopher Aristotle wrote *Meteorologica* dealing with the "study of things lifted up," including atmospheric phenomena.

Early forecasts were based on local observations made directly by the human senses. Accurate measurements of temperature and atmospheric pressure were not available until after the thermometer and the barometer were perfected in the 17th century. Comprehensive weather forecasting did not become practical until the telegraph was invented in the 19th century. This made possible the rapid collection and transmission of weather observations from many locations.

The first systematic weather observations in the United States began in 1738. One of the world's first known weather maps was drawn by the German scientist Heinrich Brandes in 1816. The next step was the establishment of a telegraphic network of observers needed to prepare daily weather maps. This came about in 1849, through the efforts of Joseph Henry of the Smithsonian Institute in Washington, D.C.

The National Weather Service has its roots in the United States government's first official weather forecasts, offered in 1870 through the efforts of the Army Signal Service. They controlled the government's telegraph system and, so, were the most likely parents. In 1891, the Army's civilian weather activities were moved to a new agency called the United States Weather Bureau, within the Department of Agriculture. In 1940, the Weather Bureau moved in with the Department of Commerce. In 1970, it was renamed the National Weather Service and placed under the Department of Commerce's National Oceanic and Atmospheric Administration (NOAA).

The Canadian Meteorological Service was set up in 1871. Exactly a century later it was renamed the Atmospheric Environment Service, an agency of the Department of the Environment.

There's even a World Meteorological Organization (WMO), established in 1951 under the United Nations. It coordinates the worldwide exchange of weather and climate information.

THE NATIONAL WEATHER SERVICE

The National Weather Service has a vast operating program. NWS personnel are found at over 400 facilities in the 50 states and elsewhere. Altogether, NWS has about 5000 full-time employees working in meteorological, hydrological, and oceanographic operations. In a single year, about 3½ million observations are taken and 2 million forecasts and warnings are issued. In addition, millions of individual briefings and services are provided on a routine but unscheduled basis to aviators, boat and ship operators, and the general public.

The National Meteorological Center (NMC) is the operating nerve center for weather information. The Center's computers incorporate more than 100,000 weather reports daily from around the world into physical and numerical models of the atmosphere. These produce weather predictions as far as ten days into the future, as well as monthly and seasonal predictions of expected temperature and precipitation conditions over North America.

Weather Service Forecast Offices (WSFOs) at 52 locations issue warnings and forecasts for over 600 zones throughout the U.S. The forecasts are issued three times a day for a 48 hour period and are updated as necessary. Extended forecasts—looking ahead five days—are issued daily for statewide areas. Local forecasts, adaptations of zone forecasts for metropolitan areas, cities, and towns, are issued by 52 Forecast Offices and more than 240 smaller Weather Service Offices.

Even with all their sophisticated computers and charts, the National Weather Service can be wrong. They just forecast the weather; they don't *make* the weather. Many times the weather observations of the human eye are just as accurate as technology—but usually not.

Let's look back at what man knew about forecasting weather before science came along.

WEATHER PROVERBS

The trouble with weather proverbs is not so much that they are all wrong, but that they are not all right for all times in all places. Some proverbs heard in New England originated thousands of years ago in northern Africa near the Mediterranean Sea, where they were heard and repeated and at last recorded by the writers of the Old Testament. Many a farmer in the Middle West, depending on a sure-fire weather saying his grandfather brought from Germany or Sweden, has found it useless in the United States. (For an accurate overview of climate in the United States, see Figs. 4-1 through 4-7.)

Distances far shorter than in either of these examples are enough to ruin some weather proverbs, those, for instance, that predict rain from the direction of the wind. When the wind blows up the side of a mountain it is cooled and loses its moisture in the form of rain; a west wind blowing up the west side of a mountain would produce the same result—a fall of rain—as an east wind blowing up the east side of the same mountain. What this adds up to is that a distance just great enough to hold a good-sized mountain might also be great enough to ruin a proverb about west (or east) winds bringing rain (Figs. 4-8 through 4-10).

Here are a few contradictory weather proverbs that were no doubt written in different places:

"Fair weather cometh out of the north." (Job)
"The north wind bringeth forth rain." (Proverbs)
"Take care not to sow in a north wind or to graft and inoculate when the wind is in the south." (Pliny)
"The north wind is best for sowing seed, the south for grafting." (Worledge, 1669)

Another point worth noting about the importance of locality is that our Pacific Coast offers moisture-bearing winds from the west and southwest, while in the East they come from over the Gulf of Mexico and the Atlantic. These two weather

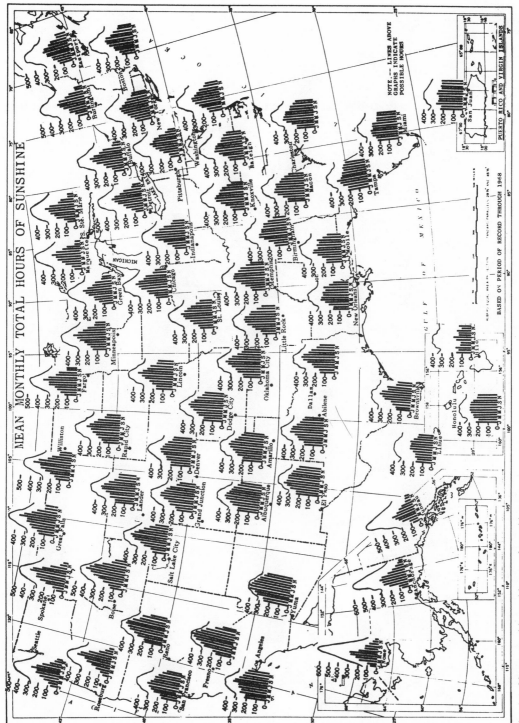

Fig. 4-1. Mean monthly total hours of sunshine.

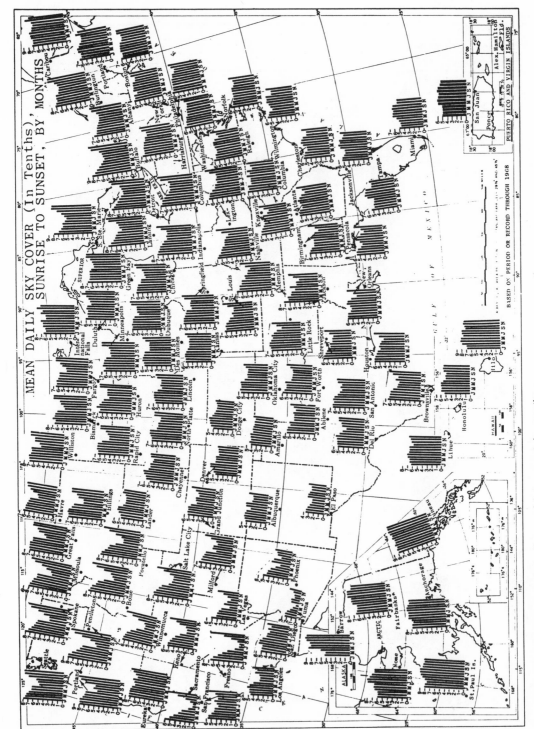

Fig. 4-2. Mean daily sky cover, in tenths, sunrise to sunset, by months.

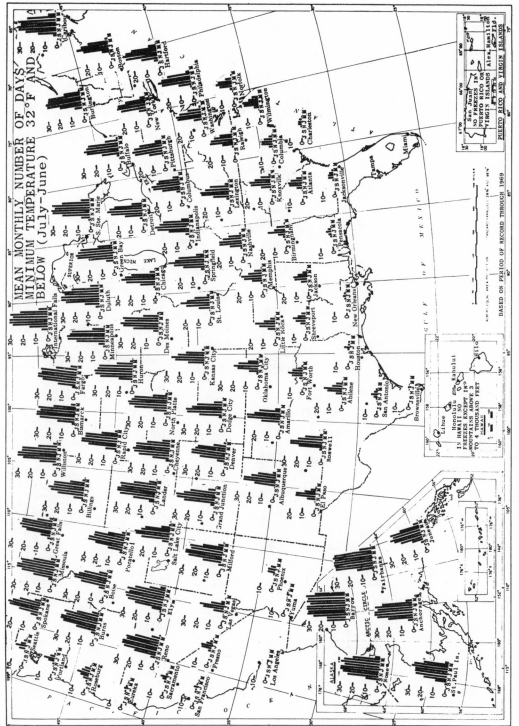

Fig. 4-3. Mean monthly number of days with a minimum temperature of 32°F and below, July through June.

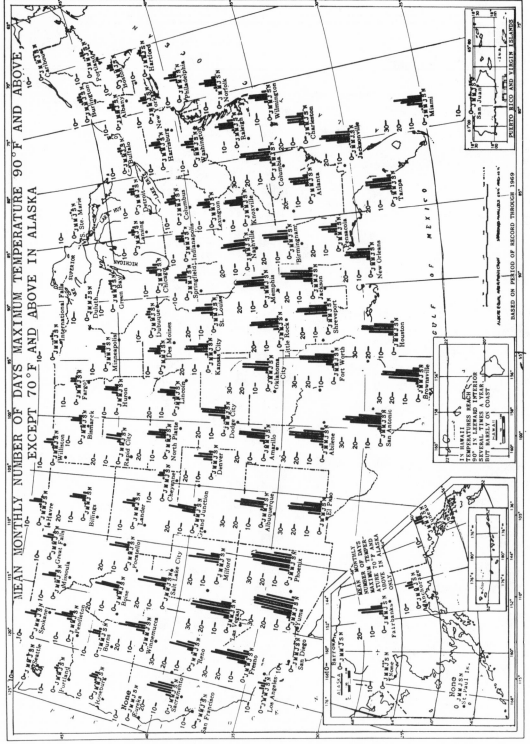

Fig. 4-4. Mean monthly number of days with maximum temperature of 90°F and above, except 70°F and above in Alaska.

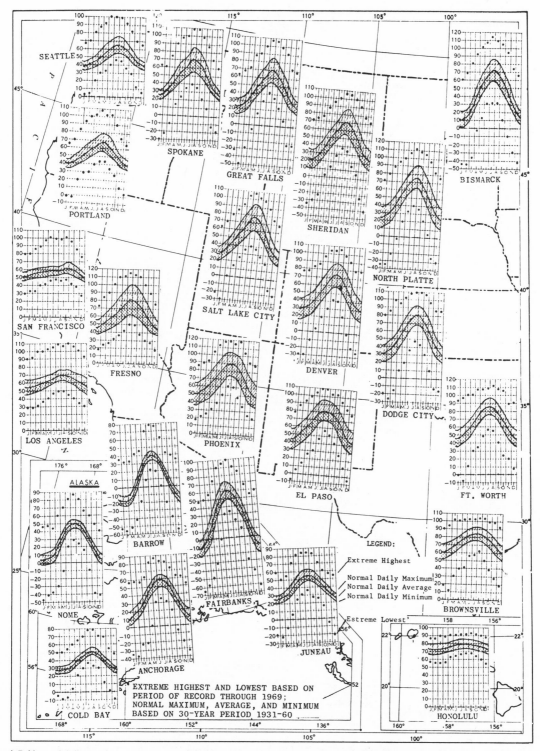

Fig. 4-5. Normal daily maximum, average, minimum, and extreme temperatures (°F) by months, for the western United States.

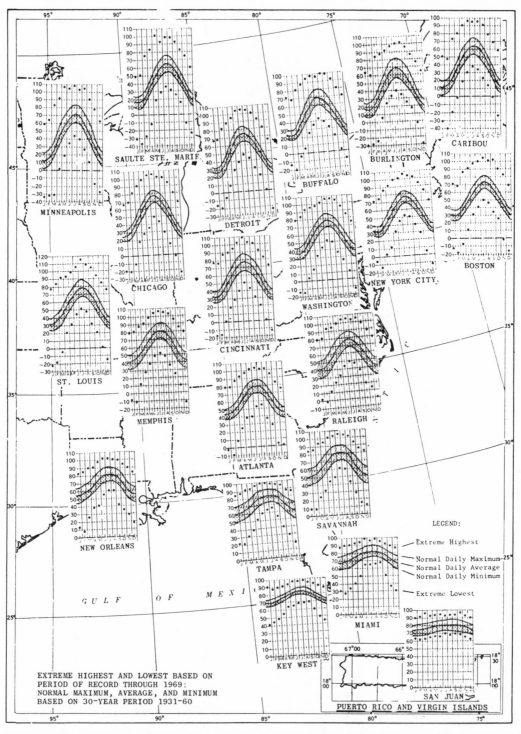

Fig. 4-6. Normal daily maximum, average, minimum, and extreme temperatures (°F) by months, eastern United States.

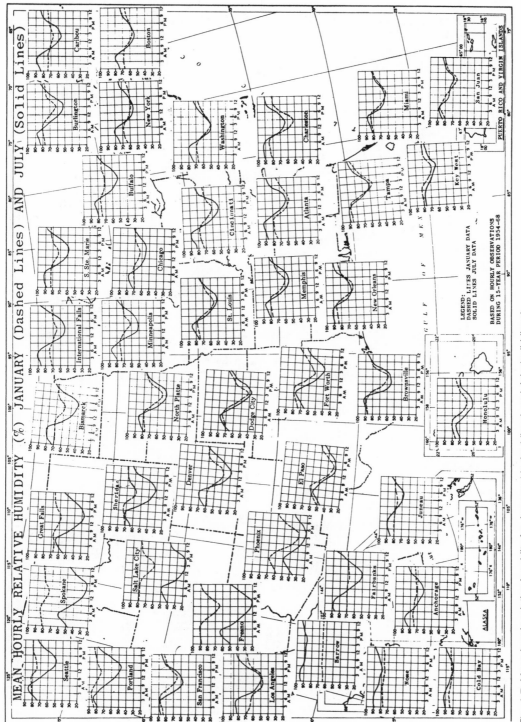

Fig. 4-7. Mean hourly relative humidity for January (dashed lines) and July (solid lines).

85

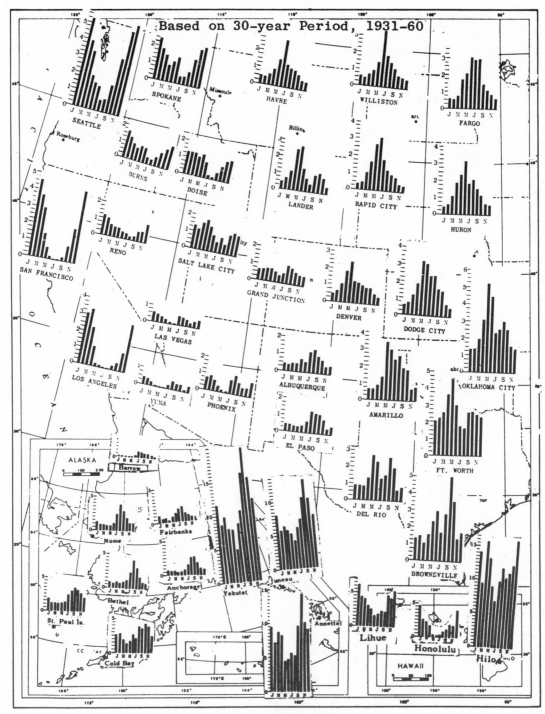

Fig. 4-8. Normal monthly total precipitation (inches), western United States.

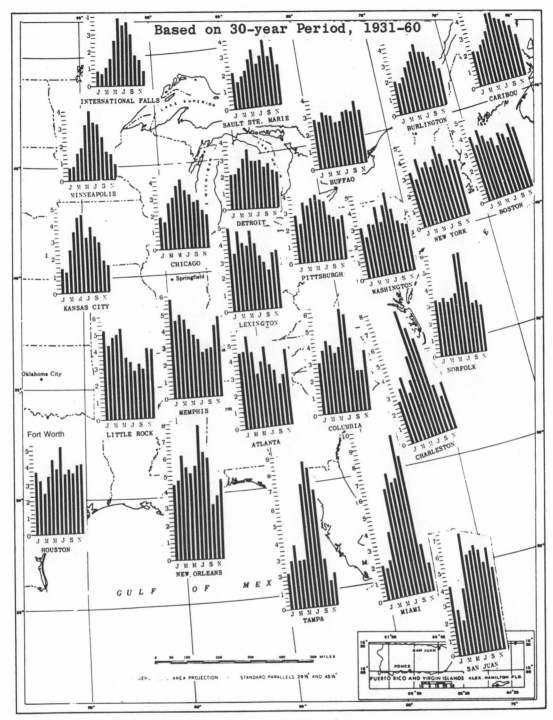

Fig. 4-9. Normal monthly total precipitation (inches), eastern United States.

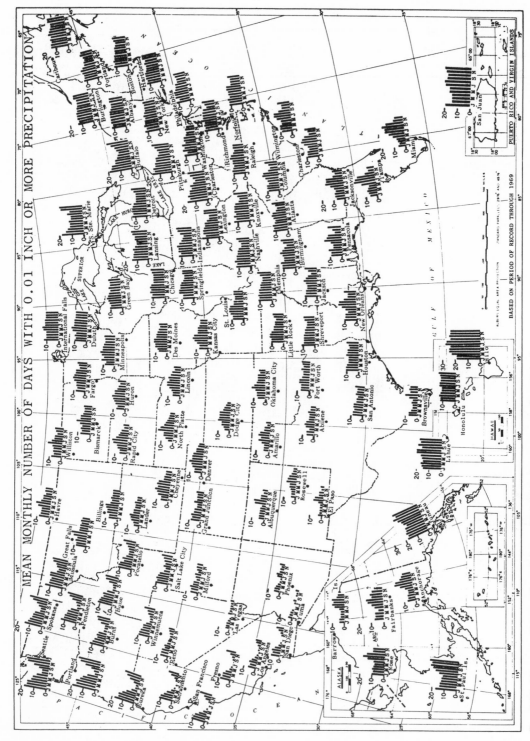

Fig. 4-10. Mean monthly number of days with 0.01 inches or more of precipitation.

proverbs, then, should not be considered too seriously in the East:

> "A western wind carrieth water in his hand."
> "When the east wind toucheth it, it shall wither."

On the other hand, few people on the west slopes of the Cascade Mountains and the Sierras where rain and snow are very frequent companions of west and southwest winds would agree with the following proverb:

> "When the wind is in the west, the weather is always best."

Proverbs concerning the southerly breezes can also be misleading, depending on where you live:

> "The south wind warms the aged," *and*

> "The south wind is the father of the poor," are hardly applicable in Louisiana. There, south winds are about the wettest, stormiest and generally least pleasant of winds in states bordering the Gulf of Mexico. The proverb writers, including Shakespeare, are noticeably consistent in pointing this out:

> "The south wind doth play the trumpet to his purposes, and by his hollow whistling in the leaves foretells a tempest and blustering sky."
> "If feet swell, the change will be to the south, and the same is a sign of a hurricane."
> "When the wind's in the south the rain's in its mouth."

Anybody who has ever looked at a collection of these sayings must have been impressed by their variety. They are extremely ancient—about as old as language itself. They illustrate the importance of weather in human affairs. They demonstrate man's hopeful opinion that experience is a good teacher. They produce some very striking relationships: wolves and crops, sky colors with foul results, holy days and unholy weather. Rain is seemingly foretold by the behavior of cats, dogs, cattle, red hair and ropes, spiders and smoke, crickets, frogs, birds, mice, flies, and rheumatism. Squirrel stores and the thickness of their fur lead to prophesies of hard winters. The drought or wetness of summers is predicted by the weather in March. What happens on Christmas foretells what will happen on Easter; light or heavy fogs in October foretell light or heavy snows in the coming winter. One proverb says "If the spring is cold and wet, then the autumn will be hot and dry;,' another "A wet fall indicates a cold and early winter," and still another "A cow year is a sad year and a bull year is a glad year."

A few other weather proverbs—good, bad, and indifferent—illustrate the variety of subject and opinion:

> "When the wind is in the south it blows the bait in the fishes' mouth."
> "One swallow does not make a summer."
> "If the weather is fine, put on your cloak; if it is wet, do as you please."
> "A bad year comes in swimming."
> "The first Sunday after Easter settles the weather for the whole summer."
> "If Candlemas Day be fair and bright, winter will have another flight; but if Candlemas Day brings clouds and rain, winter is gone and won't come again."
> "Red sky in the morning, sailors take warning."
> "Mare's tails and mackerel scales make tall ships take in their sails."
> "March comes in like a lion and goes out like a lamb."
> "Clear moon, frost soon."
> "Rain before seven, stop by eleven."
> "A year of snow, a year of plenty."
> "Rainbow in the morning gives you fair warning."
> "The bonnie moon is on her back, mend your shoes and sort your thack."
> "When the stars begin to huddle, the Earth will soon become a puddle."
> "A windy May makes a fair year."
> "Do business with men when the wind is in the northwest."

One of the best known of the rain prophesies states that 40 wet days will supposedly follow a rainy St. Swithen's Day (July 15), and the groundhog day story gets into practically every newspaper in the country during the first week in February. Since neither of these old standbys has any basis in meteorology, their persistent popularity, like that of countless others, must be explained by something else—possibly that nearly everybody on Earth since Creation has wanted to know what the weather is going to be tomorrow, next week, next month, and a year from now.

Farmers and sailors want to forecast weather because a large part of their actions and fortunes depend on the weather. But weather also affects many others: salesmen, grain speculators, washerwomen, baseball players, amusement park managers, fishermen, military leaders, and everyone else from advertisers to zookeepers.

The best explanation for the invention, persistence, and wide distribution of these weather sayings is simply that a great many of them make good sense. For example, "One would rather see a wolf in February than a peasant in his shirtsleeves," means simply that a warm February will advance the growth of vegetation so far that a subsequent hard frost will destroy it—which nobody wants, especially a farmer who depends on his crops for a livelihood. Here are three others with the same message:

"A late spring never deceives."
"Better to be bitten by a snake than to feel the sun in March."
"A wet March makes a sad harvest."

"A year of snow is a year of plenty" is just a pleasant way of pointing out that a snowy winter provides enough soil moisture to assure good crops (Figs. 4-11 and 4-12).

The familiar halo of the sun or moon is caused by the refraction of its light by ice crystals in cirrus clouds, which frequently appear when lowered air pressure and high clouds are present and rain is approaching. Thus, proverbs saying the ring around the sun (or moon) is a sign of rain—such as "the moon with a circle brings water in her beak"—are frequently right.

Several of the many signs men see in the behavior of animals and insects are worth noting, too. For example:

"A bee was never caught in a shower."

"Expect stormy weather when ants travel in lines, and fair weather when they scatter."
"When flies congregate in swarms, rain follows soon."
"Pigeons return home unusually early before rain."

The following rather inclusive one, giving several results of the low air pressure or high humidity that often precede rain should prove—if we wait long enough—that not all weather signs are wrong.

"Lamp wicks crackle, candles burn dim, soot falls down, smoke descends, walls and pavements are damp, and disagreeable odors arise from ditches and gutters before rain."

Finally, this weather proverb of dubious meteorological value:

"Dirty days hath September
April, June and November;
From January up to May
The rain it raineth every day.
All the rest have thirty-one
Without a blessed gleam of sun;
And if any of them had two and thirty,
They'd be just as wet and twice as dirty."

THE HUMAN CONDITION

One of the best ways of observing weather firsthand is by observing yourself. That's right—your body responds to changes in air pressure, humidity, and temperature. Reading the signs can give you a hint of coming weather—and can improve your health.

In fact, there's a field of medical science called *biometeorology* that studies and affirms the relationship between weather and health.

Medical studies have illustrated that people respond to almost every shift of the changing atmosphere that covers the Earth. Metabolic and chemical changes include pulse rate, body temperature, blood pressure, and urine. Individual reactions to the environment differ, just as individuals are physically different.

Sudden, severe changes in the weather have the greatest impact on people who suffer from ar-

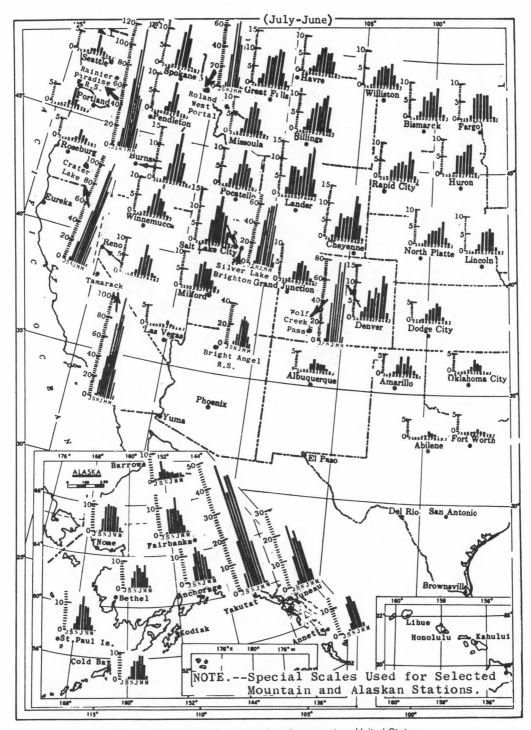

Fig. 4-11. Mean monthly snowfall (inches) for selected stations, western United States.

Fig. 4-12. Mean monthly snowfall (inches) for selected stations, eastern United States.

thritis, headaches, asthma, heart disease, or mental illness. Modern living has impaired our adaptation to the elements. Societies that live in nature—away from air conditioning, humidifiers, smog, and other unnatural conditions—adapt more easily to changing weather conditions.

Hippocrates, the "father of medicine," concluded that hot and cold winds have a bearing on the outbreak and course of disease. Some winds carry fungi spores that irritate people with asthma. Other winds mark pressure changes that irritate health.

Some people are sensitive to the electromagnetic waves that come from thunderstorms and can "feel it coming" miles away.

Increasing barometric pressure can sometimes be felt by the blood, veins, and heart, lowering oxygen levels and adversely affecting people with angina. When the barometer falls, some people feel edgy and crime rates rise slightly. Others feel exuberant.

It's been proven that moist days are healthier than dry days. Extreme dryness can cause irritation of the nose and throat, encouraging many viruses to spread. This is why doctors suggest humidifiers in drier climates.

Your build also tells you something about the weather. If you are thin—not well insulated by fat—you may be more conscious of cold because you store less energy and water. Shifting weather can more easily be felt by your body. Obese people are better insulated against winter's cold than those with heavy muscles.

Cold winters increase the chances of colds, not because of an increase of viruses as much as the increased contact with people in crowded places.

Let's consider how weather affects specific ailments.

Arthritis: The combination of rising humidity and falling barometric pressure causes arithritic joints to ache because albumin-like fluid increases the size of joint tissues and causes resistance to motion—and pain of movement.

Asthma: Doctors aren't sure why, but dramatic changes in the weather affect asthma patients. The greatest culprit is winds that carry pollens and other substances that irritate the lungs of sensitive people. The first invasion of cold air each year can also initiate asthmatic spasms.

Headaches: Atmospheric pressure changes and intense sunlight are the causes of many headaches, say doctors. As most headaches are within the sinus passages, low humidity air and sharp pressure changes are a shock to the sinus and an instigator of sinus headaches—some of them severe.

Common cold: Weather conditions don't cause colds, but they can increase your susceptibility and prolong the symptoms. The cold viruses and bacteria are always present in your body, but sharp changes in the weather can reduce your body's ability to fend them off.

Bronchitis and Emphysema: Victims of these diseases need oxygen and anything that robs the air of oxygen can make breathing more difficult and spur an attack. Weather-related causes include inversion layers over metropolitan areas that stagnate air and pollutants. High humidity also lessens the amount of oxygen in each breath and makes patients uncomfortable.

Heart attack: Research indicates that sudden changes in weather, fast-moving fronts, and changes in barometric pressure and moisture can trigger heart attacks. So can the unfamiliar exertion of shoveling new snow.

WEATHER AND THE MIND

Weather can also affect people psychologically. Moods can change, irritability can increase, crime can multiply—all because of changes in the weather.

Some people are extremely sensitive to weather and respond to every cloud and every sunny day. Their moods also reflect changes in barometric pressure and humidity. Ideal weather conditions for humans are an average temperature of 64 degrees with about 65 percent humidity, day and night. Some can adapt easily to sudden changes from this norm, but others feel uncomfortable when conditions move far afield.

Sharp changes in atmospheric pressure, studies say, trigger more aggressive behavior. A Canadian study found that most car accidents occur when the barometer is falling. A study in Tokyo found that falling pressure meant that the lost-and-found offices were unusually busy; people were more absent-minded. Suicide attempts seem to be more likely when the barometer is falling.

Another study found that assaults and related crimes increased by nearly half on warm, muggy days and nights. Conversely, seasonably comfortable weather after a period of bad weather often drops the crime rate by 75 percent.

Mental illness also increases with drastic weather changes.

So, one excellent way of observing and forecasting weather is to watch your own body react to changes in the weather. It will not only give you a clue to incoming atmospheric conditions, it will also help you understand and take advantage of physical and mental changes caused by the weather.

SKY OBSERVATIONS

You possess the greatest weather eye of all. You must see and interpret the changes in the weather around you: clouds, precipitation, winds, and nature. Here are some practical weather observations you can make to help you foretell future weather conditions.

Storm Indicators

Nature reacts to incoming storms in various ways: insects are more active, bees return to hives, bats and birds fly lower, and frogs croak more, all in response to changing atmospheric pressure and humidity.

Precipitation (rain or snow) is on the way if low clouds move in behind middle and high clouds, there is a ring around the moon (indicating high cirrus clouds), clouds begin developing vertically, a dark threatening sky is observed to the west or northwest (in most areas), or when leaves show their undersides (due to strong winds that often precede cold fronts).

Blue Skies

Positive indicators of fair weather include dark clouds becoming lighter and steadily rising in altitude (because of the passing of the cold front), a sudden wind shift (meaning the passing of the cold front), and warming winds from the direction of prevailing warm fronts in your area.

Other Weather Phenomena

Frost can often be foretold by fog forming on a pond in the spring or fall, indicating colder air temperatures that turn moist air vapor into water.

Dew is an indicator of clear weather.

KEEPING TRACK

Many people keep track of weather observations in a weather diary or as regular entries in their personal diary or journal:

"September 7, noted a red halo around the moon tonight which could mean rain tomorrow."

"The animals seem more active today. There must be a low front coming in. Watch for storm on the horizon."

"High clouds this morning should clear off by the afternoon and give us a beautiful weekend."

Once you've set up your own weather observation station with the instruments discussed in the next two chapters, you can make more specific entries in a weather log to be introduced in Chapter 7.

In the interim, you may come up with some of your own weather proverbs developed with your region in mind. Watch for the direction of incoming weather, how it is affected by mountains and bodies of water in your region, how long systems take to move from the horizon to your home, and how changing weather conditions affect your mind and your body.

You can be your own weather eye.

Chapter 5

Basic Instruments

If you've ever said, "It looks like it's going to rain," you've practiced meteorology—and, on a small scale, imitated the National Weather Service's major functions: observing, forecasting, and disseminating weather information.

A number of people, however, are more than just casual observers of the passing weather scene. Many have set up their own stations and gain a great deal of personal satisfaction from recording each day's weather and trying their hand at forecasting. These people are called *amateur meteorologists*, and you can join them. With some basic instructions and a few simple instruments, some of which can be built at home, you can be an amateur weather forecaster (Fig. 5-1).

Observing, recording, and forecasting are the present, past, and future tenses of meteorology.

Observations, taken accurately and at regular intervals, are of utmost importance to the weather forecaster. Weather changes are not usually heralded definitely by local indications for periods longer than a few hours in advance. Indeed, many local storms give scarcely an hour's notice of their coming.

Records kept of past weather can be a valuable tool in predicting future weather. The rules in this book for forecasting weather from local indications are usually reliable, but may not apply in all locations and situations. To form your own rules for your area you must carefully and systematically record and correlate your own observations.

Forecasting is the real challenge of meteorology. Although there has been much exaggeration of the ability of mariners to forecast weather changes from local observations, they are somewhat more adept than people in most other occupations. Pilots and farmers who depend more on the weather are exceptions, too. This is not because signs are more pronounced over the ocean, in the flier's sky, or over farms, but primarily because these people have learned to interpret and forecast the changes in weather conditions.

THE WEATHER SHACK

The first thing you'll need at your weather station is a weather shack, a small shelter for the instruments. For best results, the various thermometers needed should be kept outdoors in a

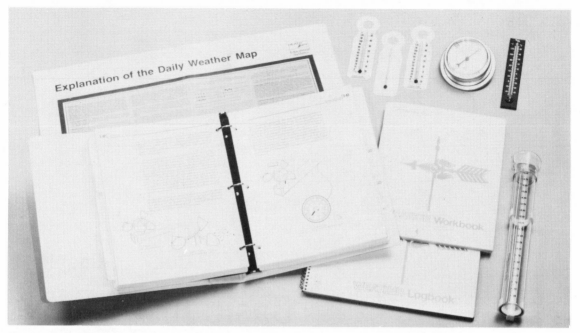

Fig. 5-1. Basic weather instruments are important to studying meterology. (Courtesy Heath Company)

shelter. A thermometer really only takes its own temperature and if this is to represent the air temperature the thermometer must be exposed in such a way that it is subject to the same temperature as the air. This means keeping it out of the sun and keeping it where the air can move freely around it. The best way is to put it in a ventilated box or thermometer shelter.

A simple shelter can be built from a couple of window shutters, some wood or beaver board, and a supporting post or posts (Fig. 5-2). The shutters form the sides of the shelter. The back, top, and bottom can be made from two layers of wood with a small air space in between. It can have a hinged door or an open side, but the opening should face north to keep the sun from shining in and affecting the thermometers when the door is open. The shelter should be painted white, inside and out.

WIND INSTRUMENTS

Wind is air in motion. As air in motion, the wind has four important properties of vital interest to the meteorologist: direction, speed, character, and shifts. The *character* of the wind refers to its

Fig. 5-2. The standard instrument shelter or "weather shack;" (A) how it's built and (B) what it contains.

gustiness while the *shifts* of the wind refer to its steadiness or unsteadiness in direction.

More specifically, *wind direction* is the direction *from* which the wind is blowing. It is usually expressed in compass points—northeast, southwest, east—or in degrees from north—180 (south), 230 (southwest), etc.

Wind speed is the rate of motion of the air in a unit of time. Wind speed can therefore be measured in knots (nautical miles per hour) or mph (miles per hour).

Gust is defined as rapid fluctuation in wind speed with a variation of 10 knots or more between peaks and lulls. *Squalls* are sudden increases in wind speed of at least 15 knots sustained at 20 knots or more for at least one minute. *Peak gust* is the highest instantaneous wind speed observed or recorded.

Wind shift is a change in wind direction of 45 degrees or more that takes place in less than 15 minutes. Wind shifts are normally associated with the passage of a cold front. Wind shifts foretell of a rapid drop in the temperature and dew point, a rapid rise in pressure, and a coming storm.

Wind Vane

Wind direction can be determined from the flow of smoke, the behavior of a flag, or even the time-honored (if somewhat unscientific) wet-finger method. If a wind vane is in sight, so much the better.

A wind vane can be easily fashioned from wood, tin, or other material. Both ends should be balanced and it should swing freely on a central pivot. Ornamental wind vanes may be purchased at small cost, adding a decorative value to their scientific usefulness.

The vane should be mounted where the wind is least affected by local influences such as trees, buildings, and windbreaks. The direction indicators should be carefully lined up with cardinal compass points.

A wind sock can easily be constructed to give you both wind direction and speed (roughly). A square yard of light-weight material can be cut in a wide funnel shape, 36 inches wide narrowing down

to 12 inches. The fabric is then sewn together to form a wind funnel and attached to a piece of heavy wire or a coathanger formed in a circle. Curtain rings will keep the sock loosely attached to the frame. A cardboard circle can then be cut out, marked with the compass points, and slid onto the broomstick or garden stake that will hold the windsock in place. A compass can align the cardboard marker. Mounted clear of obstructions, your wind sock can illustrate the direction from which the wind blows as well as estimated speed: limp means calm, straight out means about 30 knots (35 mph) and halfway between indicates a wind of about 15 knots (17 mph).

Anemometer

To measure wind speed more accurately you'll need an *anemometer*. They can be built at home, but few are as satisfactory as the gearbox of a cup-type anemometer. One relatively useful homemade anemometer is shaped somewhat like a slingshot with a flap suspended between its arms which swings back along a scale as the wind strikes it. This instrument is calibrated by holding it out of the window of a moving car and marking the point to which the flap is blown back along the scale at different speeds.

As with the wind vane, readings with this instrument should be made where the wind is least affected by local influences. The anemometer, of course, should be headed directly into the wind while a reading is being made.

BAROMETER

While the barometer has been greatly overrated as a forecasting device, it is essential that the local observer be equipped with some means for detecting changes in atmospheric pressure. That's the job of the barometer. Pressure changes used in connection with the wind direction will give the best key to the local weather situation (Fig. 5-3).

The barometer is, by far, the most expensive instrument the amateur meteorologist may have to purchase. Homemade barometers are rarely accurate enough for serious weather observation. If your budget is short, you can use a weather radio

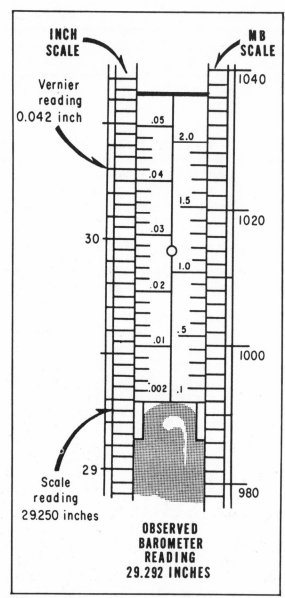

INCH SCALE

MB SCALE

Vernier reading 0.042 inch

1040

.05

2.0

.04

1.5

1020

.03

30

1.0

.02

.5

.01

1000

.002   .1

980

29

Scale reading 29.250 inches

OBSERVED BAROMETER READING 29.292 INCHES

Fig. 5-3. Reading the mercurial barometer in inches of mercury and millibars.

(Chapter 7) to learn the latest barometric pressure in your area. Otherwise, a serviceable aneroid barometer can be purchased for about $30.

The barometer should not be placed out-of-doors because such exposure may corrode its parts. Since air pressure is the same both indoors and outdoors, the instrument can be located inside at

some convenient spot. Keep it out of sunlight and away from drafts, in a place where the temperature does not vary too much and the readings will be reliable.

The barometer should be adjusted to sea level. Instructions for such adjustments will be included with the barometer. The current sea level pressure is available from local Weather Service facilities.

Of the instruments combined in thermometer/barometer/humidity instrument sets, the barometer will be the only aid to the amateur meteorologist. Temperature and humidity readings taken indoors are worthless in forecast work and merely indicate atmospheric conditions within a particular room of the house. Better instrument sets remotely read outside temperatures and humidity.

To help you in shopping for the best barometer for your weather station, let's look at the types, how they work, and what they do.

Mercury Barometers

Nearly three and a half centuries ago, an Italian physicist named Torricelli made the first crude barometer. It was made with a long glass tube, open at one end, closed at the other, and filled with mercury. The open end was sealed temporarily and placed into a basin of mercury, then unsealed (Fig. 5-4). This allowed the mercury in the tube to descend, leaving a nearly perfect vacuum at the top of the closed end of the tube.

When the atmospheric pressure is increased, the mercury in the basin or cistern is forced into the glass tube. As the atmospheric pressure is decreased, the mercury in the tube flows into the cistern. The height of the mercury column in the tube is therefore a measure of the air pressure. This is where we get the pressure measurement term "inches of mercury."

Aneroid Barometer

Mercury barometers are quite accurate, but they are expensive and not easy to move around, being nearly three feet high. For most purposes they are replaced or supplemented by a mechanical instrument known as the *aneroid barometer*. "An-

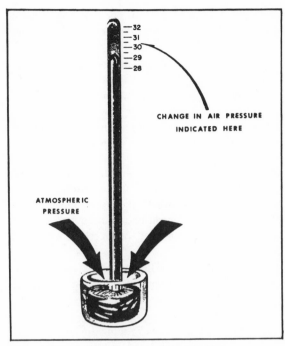

Fig. 5-4. Principle of the mercurial barometer.

eroid" means "without fluid." The aneroid barometer, then, is a fluidless barometer, utilizing the change in shape of an evacuated metal cell to measure variations in atmospheric pressure.

The aneroid barometer gets its name from the pressure-sensitive element used in the instrument. It is an aneroid, which is a thin-walled metal capsule or cell, sometimes called a diaphragm, that has been either partially or completely evacuated of air. The aneroid is usually made of beryllium copper or phosphor bronze. Most aneroid cells are self-supporting and do not require external or internal springs to prevent the crushing of the cell walls by atmospheric pressure.

In a common type of single aneroid cell barometer (Fig. 5-5), the top of the evacuated cell is secured to a suitable linkage which transmits the motion of the aneroid to an index hand or pointer that indicates the pressure on the face of the instrument.

If you just fell into some money you may want to buy a precision aneroid barometer with highly

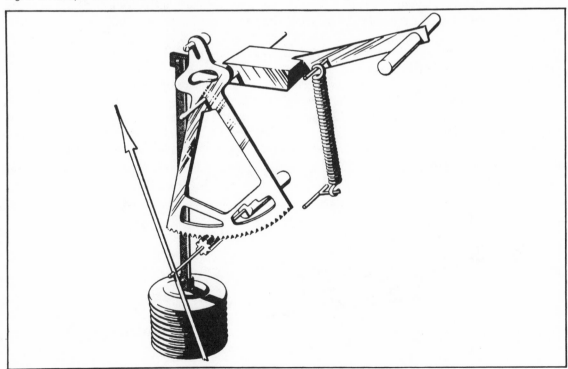

Fig. 5-5. Simple diagram of the aneroid barometer.

accurate readings. They compensate for temperature changes and use a Sylphon cell for a more sensitive reading of the atmospheric pressure.

The ultimate barometer is a computerized barometer that uses an electronic transducer to read atmospheric pressure rather than an aneroid diaphragm. One such model sells for under $200 and offers current pressure, rate of change, direction of change, maximum and minimum pressures from memory, as well as the date and time they occurred. It is also a clock. Readings are in inches of mercury, millibars, or kiloPascals (ten millibars).

If you're an aviator you know that *altimeters*, which measure a plane's altitude above sea level are simply aneroid barometers calibrated to indicate altitude in feet rather than units of atmospheric pressure—same innards.

THERMOMETERS

The thermometer is another important weather instrument that can be built by the amateur or bought for a few dollars.

If maximum and minimum temperature records are to be kept, a special thermometer, or set of thermometers, will be needed.

The National Weather Service uses a set of two thermometers to obtain maximum/minimum readings. The maximum thermometer has a special constriction near the bulb and the minimum thermometer has a floating index. Cost for the professional model runs $40 or more.

A less expensive, longtime favorite of the amateur forecaster is called the Six's-type, named after its inventor. This U-shaped thermometer with one side reading maximum and one side reading minimum temperatures can be bought for less than $20.

A dial-type maximum/minimum thermometer is also available. This type has a coiled metal spring temperature element and two thin metal pointers that are pushed either up or down into position as the main point of the temperature element moves around the scale.

Thermometers, of course, should be located outdoors and in constant shade. A properly constructed and situated instrument shelter is ideal for thermometer exposure, as discussed earlier. However, thermometers can be mounted on the shady side of unheated buildings, or on trees, with accurate results.

The standard air thermometer that you are more than likely familiar with (Fig. 5-6) is the one placed inside or outside the house to see how cold or warm the temperature is during the day or how cool the air conditioner is keeping the house. There are many uses for this thermometer, which is usually filled with mercury or alcohol, depending on its intended use. These two fluids are used because they have a much greater coefficient of expansion for each degree of change in temperature than the glass in which they are housed (Fig. 5-7).

The standard air thermometer used by most stations has a range from −20F to +120F.

Handle these thermometers carefully because they break easily. It is important that the thermometer stem and bulb be kept clean and free of dirt, dust, and moisture since the presence of these elements can cause errors in free air temperature

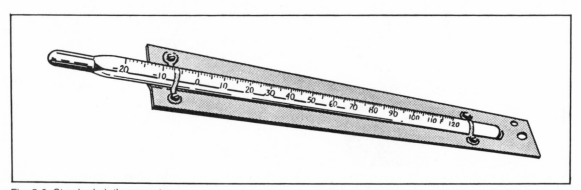

Fig. 5-6. Standard air thermometer.

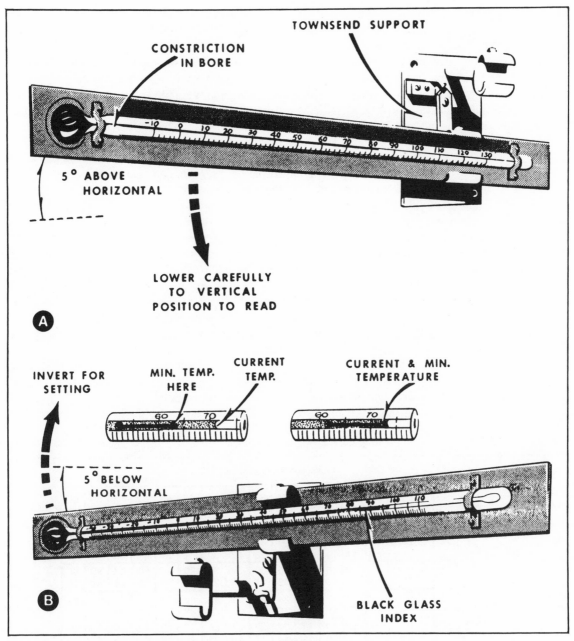

CONSTRICTION IN BORE

TOWNSEND SUPPORT

5° ABOVE HORIZONTAL

LOWER CAREFULLY TO VERTICAL POSITION TO READ

A

INVERT FOR SETTING

MIN. TEMP. HERE

CURRENT TEMP.

CURRENT & MIN. TEMPERATURE

5° BELOW HORIZONTAL

BLACK GLASS INDEX

B

Fig. 5-7. Maximum thermometer, top, and minimum thermometer, bottom.

readings. Clean the stem and bulb by wiping with a soft cloth. This should be done ten to 15 minutes prior to taking a reading so the temperature will have time to stabilize before the observation. Remove and clean the metal back as necessary. Upon reassembly apply a drop of light oil to any mounting screws, if needed. Renew the etched graduations when faded.

To reunite a separated mercury column, tightly attach a string to the thermometer back and whirl it, or tap the bulb lightly against the heel of the fleshy part of the hand to jar the mercury column back together. If this fails, gently heat the bulb by placing it near a light bulb until the column unites. *Never* heat the bulb over an open flame. Leave a small space at the top of the tube while heating; otherwise the thermometer will break. If these methods fail to unite the mercury column, the thermometer should be replaced.

HYGROMETERS

There are six different means of measuring the water vapor content (humidity) of the atmosphere and hence an equal number of types of hygrometers. The *psychrometer* is the simplest type and can be built at home without too much trouble, or purchased at a reasonable cost.

The psychrometer consists of two matching thermometers, one, exposed to the free air, is the dry bulb and the other has its bulb covered by a water-saturated wick, and is called the wet bulb. The wet bulb measures the temperature at which water is evaporating from the wick. Since the rate of evaporation is controlled by the amount of moisture in the air and evaporation is a cooling process, the wet bulb will read lower than the dry bulb. This difference in temperature is known as the wet bulb *depression*. When this depression figure is determined by subtracting the wet bulb temperature from the dry bulb temperature, Table 5-1 can be used to find out the relative humidity.

A simple wet bulb thermometer can be made from a regular thermometer, a bootlace, and a small medicine bottle. It's best to boil the bootlace before using it to get all the impurities and coloring matter out of it. Use about three inches of a tubular white cotton bootlace. Slip one end of it over the bulb of the thermometer. (Part of the thermometer's wooden or metal backing may have to be cut off.) Arrange the thermometer so that the other end of the bootlace dips into a small bottle filled with water. The water will soak up into the bootlace and keep the bulb moist. The wet bulb thermometer may then be mounted on a suitable backing with another thermometer to form a psychrometer. *Note*: Make sure the temperature reading on the two thermometers is the same before they are purchased.

A factory-built psychrometer (Fig. 5-8), built to Weather Service specifications, can be purchased for around $30.

Hygrometers should be kept out-of-doors in a

Table 5-1. Relative Humidity Table.

Wet bulb temperatures

Dry bulb	45	46	47	48	49	50	51	52	53	54	55	56	57	58	59	60
70	−14	0	+9	16	21	26	30	33	36	39	42	45	47	49	51	53
	3	6	9	12	16	19	22	26	29	33	36	40	44	48	51	55
71	−26	−5	+5	13	19	24	28	32	35	38	41	44	46	48	51	53
	2	5	8	11	14	17	20	23	27	30	34	37	41	45	48	52
72		−13	+1	10	16	22	26	30	34	37	40	43	45	47	50	52
		3	6	9	12	15	18	21	24	29	31	35	38	42	45	49
73		−26	−5	+6	13	19	24	29	32	36	39	41	44	47	49	51
		2	4	7	10	13	16	19	22	25	29	32	35	39	42	46
74			−13	+1	10	17	22	29	31	34	37	40	43	46	48	50
			3	6	8	11	14	17	20	23	26	30	33	36	40	43
75			−25	−4	+6	14	20	25	29	33	36	39	42	45	47	49
			2	4	7	10	12	15	18	21	24	27	31	34	37	40
76			−57	−12	+2	11	17	23	27	31	35	38	41	44	46	49
				3	5	8	11	14	16	19	22	25	28	31	35	38
77				−23	−3	+7	15	21	25	30	38	37	40	43	45	48
				2	4	7	9	12	15	18	20	23	26	29	32	35

Pressure 30″

Elevation 0-500

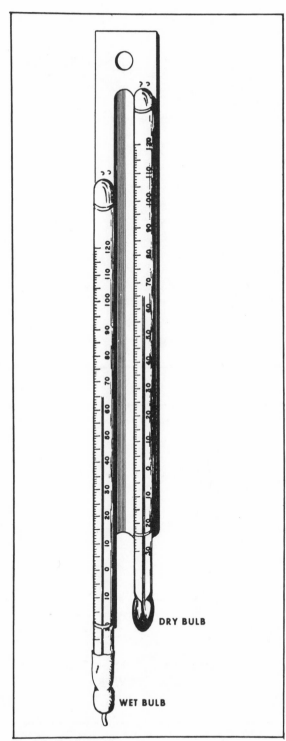

DRY BULB

WET BULB

Fig. 5-8. Psychrometer.

properly ventilated shelter. In below-freezing temperatures, care should be taken outside between readings.

Sling Psychrometer

The sling psychrometer is shown in Fig. 5-9. The sling consists of a wooden grip with a swivel head and harness-type snap or spring clip for attaching to the top hole of the psychrometer frame.

When not in use, the sling psychrometer should be hung on a suitable hook. Handle the sling psychrometer carefully at all times. The thermometers are easily broken through careless handling, dropping, or striking some object while being whirled.

For a check on humidity, take it to a clear and open place, preferably exposed to the wind. Never touch the bulb or stem in handling or expose it to the direct rays of the sun while making an observation. The bulb of the wet bulb thermometer, which is covered with a wick, is moistened with clean water at the time an observation is made. Stand in a clear shady place facing into the wind and hold the pyschrometer as far in front of the body as possible. Rotate the psychrometer with the wrist (Fig. 5-10). Bring the psychrometer to a stop without any sharp jar and bring to eye level. Then read both thermometers to the nearest tenth of a degree, reading the wet bulb thermometer first. The whirling is repeated and other readings are made until two successive wet bulb readings are the same.

Rotor Psychrometer

The rotor psychrometer shown in Fig. 5-11 is a psychrometer element secured to a handcrank-operated shaft. The rotor is designed for wall mounting and permits a permanent installation in the side wall of the instrument shelter. The same procedure is followed in obtaining a correct reading of the rotor-mounted psychrometer as was given for the sling psychrometer, except that the instrument is left in the shelter.

Hand electric psychrometers are also used by both amateur and professional meterologists (Fig. 5-12).

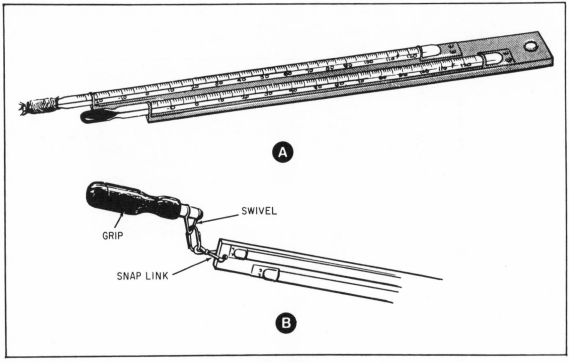

Fig. 5-9. Standard psychrometer, top; with sling attached, bottom.

RAIN GAUGES

Any straight-sided container may be used for making an inexpensive rain gauge. The rainfall for any given period of time is the depth of the rain falling on a horizontal surface during the period considered. If, therefore, the container has straight sides and the same area of cross section as the container opening and is exposed in a horizontal position, the depth of catch measured in inches and tenths may be taken as the correct rainfall value for location.

A common #10 can (6 1/16 inches inside diameter by 7 inches tall) will do very well for an improvised rain gauge. The can should be exposed in an unsheltered place, and a means provided to keep it level and protected from upsetting in the wind. It would be preferable to have the top edge of the can rolled to reinforce it so it will hold its circular shape.

The depth of the water can then be measured by means of a wooden ruler marked in inches and tenths. If a more accurate measurement is desired,

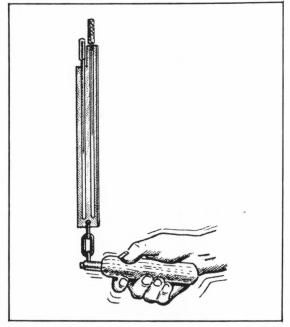

Fig. 5-10. The sling psychrometer is swung in a circular motion before the reading.

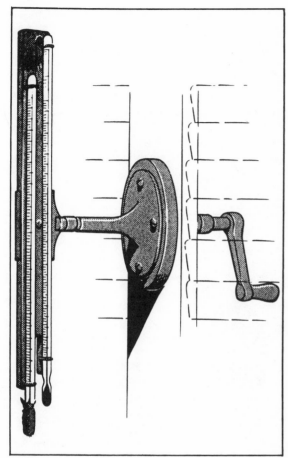

Fig. 5-11. Rotor psychrometer.

over the top of a tripod-mounted overflow container and a clear plastic measuring tube (Fig. 5-13). The rainfall drains from the collector ring through the funnel into the measuring tube, which has graduations in the wall for direct reading to the nearest hundredth of an inch.

The only maintenance required for the standard 4-inch rain gauge is to keep it clean at all times and make sure it is mounted firmly.

MEASURING CLOUDS

Clouds, cloud ceilings, and visibility are important to the weather forecaster—but are especially important to those who depend on visibility, such as aircraft pilots. Some of these observations require advanced instruments like those in the next chapter. But simple cloud observations can also be made with the human eye and mind.

First, a few terms should be defined. Clouds, as you remember, are a visible collection of minute

the water may be poured from the #10 can into a #303 can, which happens to have an area of cross section about one-fourth that of the larger can. Thus the value obtained by measuring the precipitation when poured into the smaller can divided by four will be more accurate than if measured directly in the large can. (Canned fruits and vegetables often have the can size printed on the label.)

An exposure site should be chosen which is a few feet from the ground and well away from buildings and other obstructions. Care should be taken to see that the rain does not splash into the can from the support post or any other nearby object.

The non-recording plastic rain gauge used by many amateur observation stations consists of a 4-inch diameter collector ring and funnel, which fit

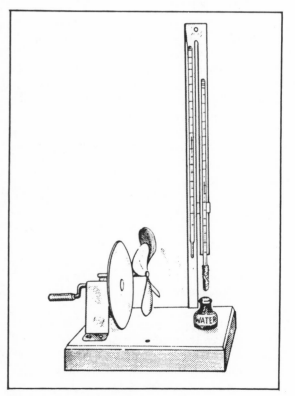

Fig. 5-12. Fan psychrometer.

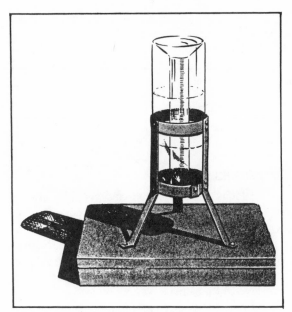

Fig. 5-13. Standard 4-inch rain gauge.

used to measure the distance from the spot to the instrument. Using the baseline (Fig. 5-15), the height of the cloud base can then be figured. Most ceiling measurement instruments are fancy variations on this principle.

Visibility is measured by an instrument called the *transmissometer.* It resembles an electric eye that reads out in variable figures to illustrate the degree of visibility.

When estimating visibility with the naked eye and pre-selected landmarks you can rate the visibility as shown on page 107.

particles of water, ice, or both in the free air. Clouds may also contain some foreign particles such as dust or smoke. *Layer* or cloud layer is the equal height of a large group of clouds (such as a cloud layer at 4000 feet). An *obstruction* means the observer cannot see more than 10 percent of the sky. *Partial obstruction* then must mean that some of the sky can be seen, measured in tenths. The *ceiling* is the height of the lowest opaque layer of clouds. A low ceiling means that clouds are low, while an unlimited ceiling means that there are no opaque clouds. Finally, *visibility* is defined as the greatest distance at which selected objects can be seen and identified. Visibility is decided by measuring the distance from the observation point to prominent landmarks at varied distances during a clear day. Then, when the fog comes rolling in, you can accurately record that "visibility is one mile" by seeing the church steeple a mile away, but none of the farther objects (Fig. 5-14).

The simplest ceiling measurement device is the *clinometer,* the portable hand instrument used to measure the angular elevation of a projected light "spot" on the base of a cloud. A projector is mounted on the ground to project a light beam on to the bottom of a cloud above it. Then the clinometer is

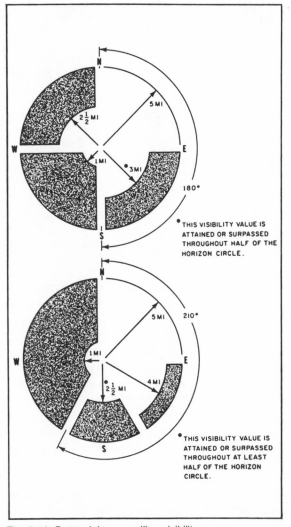

Fig. 5-14. Determining prevailing visibility.

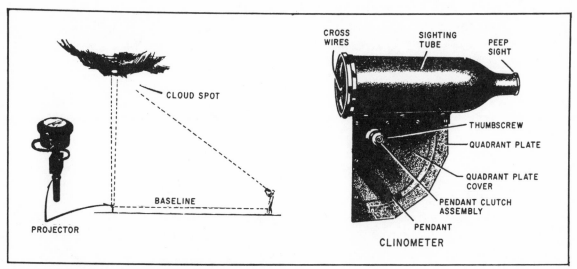

Fig. 5-15. Ceiling light projector and clinometer.

Distance	Visibility
200 yards	fog
1000 yards	mist or haze
1 mile	poor
5 miles	moderate
over 5 miles	good

Simple visibility measurements such as this can be made two or three times a day, such as 8 a.m., noon, and 4 p.m.

With the human eye and a few inexpensive or do-it-yourself weather instruments you can make accurate and useful observations of the atmosphere around you—the first important step to forecasting tomorrow's weather.

Chapter 6

Advanced Instruments

A variety of advanced and specialized instruments are used by amateur and professional meteorologists around the world. Though many may be beyond your current budget, they can often be purchased as government or military surplus at reasonable prices. Some are offered to help you understand what services are available *free of charge* through the National Weather Service. Still other instruments can be economically justified if applied to a commercial venture such as a farm, radio station, newspaper, or business that depends upon the weather (Fig. 6-1).

In any case, it's fun to see how the state of the art has advanced in the fascinating world of meteorology.

CATCHING THE WIND

The wind measuring set used by many professional weather observers is designed to provide a visual indication and a printed record of wind direction and speed values. The set includes a transmitter, a support, and a recorder. Some also use an indicator. The set is shown in Fig. 6-2.

The transmitter is a vane mounted on a vertical support. The tail of the vane brings the nose into the wind. The nose consists primarily of a screw-type impeller directly coupled to a tachometer-magneto. The magneto voltage output is directly proportional to the wind speed and is connected to the plug in the transmitter's vertical support through brushes and sliprings, then down to the indicator or recorder whose voltmeter automatically indicates or records the voltage in knots (nautical miles per hour). Motion of the vane is transmitted mechanically to the synchro located inside the enlarged section of the vertical support.

The transmitter is placed on top of a connector housing. The electrical cable, leading from the housing through the support, goes to any one or all of any combination of six repeaters. A follower synchro then converts the electrical energy into wind direction indication or recording.

Electronics technology has reduced the cost and increased the efficiency of weather instruments available to the amateur. For example, the digital wind speed/direction indicator (Fig. 6-3) offers an

Fig. 6-1. Many people have remote weather stations in their homes offering barometric pressure, wind direction, and speed as well as outdoor temperature. (courtesy Heath Company)

outdoor transmitting unit, anemometer, and weather vane made of weatherproof material and an indoor unit with seven-segment digital display offering the wind speed while a compass rose lights up to show the direction of the wind. The instrument is available in simple kit form for about $100.

THE BAROGRAPH

The barograph is a *baro*meter measuring air pressure and recording readings on a *graph*. The barograph is a highly accurate and sensitive (translated: expensive) instrument used to keep track of levels and changes in barometric pressure for forecasting purposes. Figure 6-4 shows a common barograph.

There are usually three principal sections in the barograph: the chart drive assembly, the element assembly, and the pen shaft assembly. The last two assemblies are of less interest to the barograph user, who will change the chart paper every two to 31 days depending on the model used.

The chart drive assembly consists of a chart drive mechanism, a chart cylinder, the chart, and a clip. The chart is turned either by a spring-wound

clock or by electricity—again, depending on the model.

The barograph works simply. A fine pen or etcher is attached to the top of an aneroid barometer cell and the other end lightly touches a revolving cylinder that has graph paper mounted on it. The changes in air pressure are recorded permanently on paper, based on time. Both the reading (in millibars or inches of mercury) and the relative changes (rising, falling) are available for study.

Digital barographs are also available. Figure 6-5 shows a digital barograph available in simple kit form for less than $300. It provides digital readings of the current barometric pressure and records pressure changes permanently in inches of mercury, millibars, or kiloPascals on a drum chart that can be adjusted to record at 7-day and 31-day speeds.

This model offers a memory that can recall the maximum and minimum barometric pressure readings since last cleared and the date and time of the readings. A separate display alternately shows the current time and date.

The manufacturer claims accuracy comparable

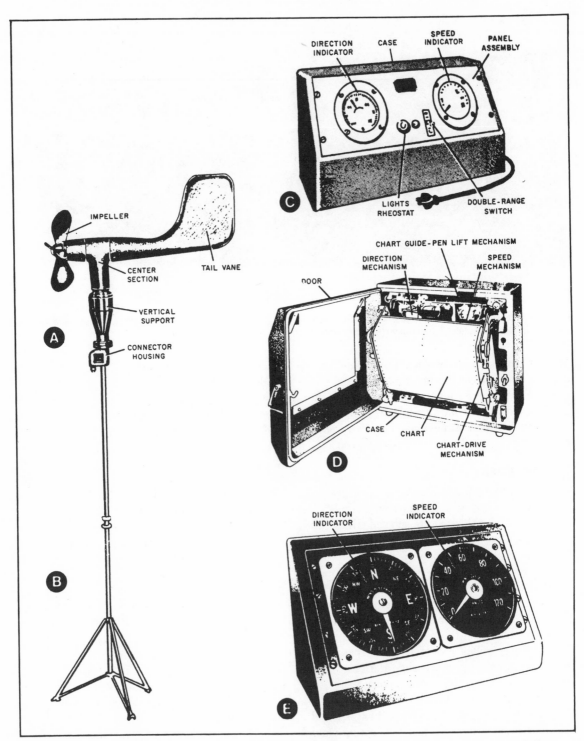

Fig. 6-2. Wind measuring set includes (A) transmitter, (B) support, (C) indicator, (D) recorder, and (E) remote indicator.

Fig. 6-3. Digital wind speed and direction indicator. (courtesy Heath Company)

Fig. 6-4. Barograph. (courtesy Heath Company)

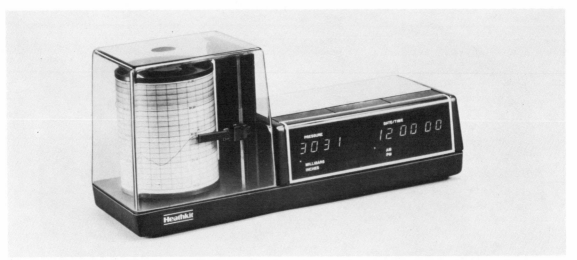

Fig. 6-5. Digital barograph. (courtesy Heath Company)

to a mercury column barometer due to a temperature-controlled oven to minimize the effects of temperature changes on the instrument.

THE THERMOGRAPH

Even the simple thermometer has been updated with the *thermograph* that not only reads the current air temperature, but also records it on a revolving chart similar to the barograph just discussed. The thermograph uses a heat-sensitive metal coil to move a pen against the rotating drum.

Figure 6-6 shows a digital indoor/outdoor thermometer instrument available for less than $100 in kit form. It can be assembled with a soldering iron or gun and conventional hand tools. One sensor reads indoor temperature and the other reads outdoor (or any other remote) temperature. A switch on the back permits setting the display for a

Fig. 6-6. Digital thermometer. (courtesy Heath Company)

Fig. 6-7. Hygrothermograph.

constant reading of either temperature, or to alternate at four second intervals between the two. A second switch lets the user select either Fahrenheit or Centigrade (Celsius) display.

HYGROGRAPH

Figure 6-7 shows another useful graph-type weather instrument, the *hygrograph* ("moisture recorder"). In this instrument, several strands of human hair are anchored at one end and connected through a system of cams and levers to a movable pen at the other. The pen rests on a rotating, chart-covered drum driven by clockwork or electricity. As the relative humidity varies, the hairs change in length, becoming longer as the humidity increases and shorter as it decreases. This change in length is recorded on the chart by a pen, giving a continuous

record of the relative humidity. The chart is calibrated so that the humidity can be read directly. Some types of instruments also have a bimetallic temperature sensing unit that changes in shape as the temperature changes. This movement is recorded by another pen on the clock-driven drum, thus providing a temperature record. When both temperature and humidity are recorded, the instrument is called a *hygrothermograph*—a good word to remember for Scrabble®.

RECORDING RAIN GAUGE

The amount of rain falling can be recorded by using what's called a *tipping bucket rain gauge.* The tipping bucket rain gauge is mounted in a housing which permits rain to fall directly on the gauge. The instrument is a two-compartment container that pivots within a casting. Rainfall enters through the upper funnel in the housing into one compartment of the bucket until 0.01 (1/100) of an inch of rainfall has accumulated. The weight of this amount of rain unbalances the bucket, causing the unit to tip on its pivots, dumping the accumulated rainwater and moving the other compartment directly under the funnel.

When the bucket tips, its rainfall content falls into a funnel beneath the bucket. At the base of the funnel is a drain cock, which in its closed position permits the rain to collect so that it may be drained

Fig. 6-8. Digital rain gauge. (courtesy Heath Company)

into the cylinder below the funnel at the time of measurement. If no purpose for the collected rain exists, the drain cock may be left open and the cylinder removed.

The tipping motion of the bucket actuates a mercury switch in the casting. Momentary contact is established within the switch, causing an electrical impulse to be sent to a recorder which registers each "count" of 0.01 inch of rain.

A digital readout rain gauge can be purchased, with automatic-dump bucket, for less than $200. The collector is mounted on a fence post, TV tower, or in an open area with a protective screen to keep out dirt, leaves, and other debris. The remote unit reads both short- and long-term rainfall in either centimeters or inches on an LED display (Fig. 6-8).

MEASURING CEILINGS

As mentioned in the last chapter, the height of cloud bases are important to the weatherman, -- but is especially vital to fliers. VFR (visual flight rules) pilots cannot legally fly if clouds are too low and/or too dense.

The ceiling light projector is a weatherproof drum housing a light and a reflector designed to cast strong light on the bottom of clouds so they can be read with the clinometer to determine the cloud height (Figs. 6-9 through 6-11).

Weather stations at airports often use a group of instruments called the *cloud height set* (Fig. 6-12) to both indicate and record ceiling height for use in aviation weather forecasts. The parts to this set include a projector, detector, indicator, and re-

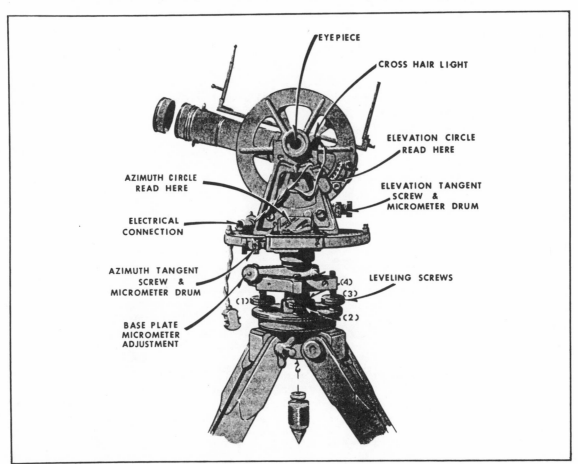

Fig. 6-9. Theodolite.

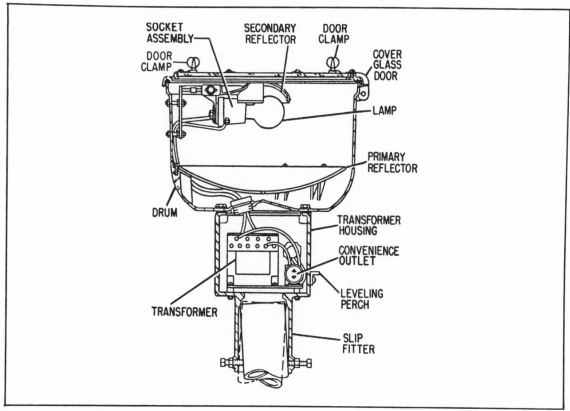

Fig. 6-10. Ceiling light projector.

corder. The recorder draws lines on graph paper indicating the height AGL (above ground level) of cloud bases as well as the density of cloud cover.

AUTOMATIC WEATHER STATIONS

Figure 6-13 shows the automatic weather station used by professional forecasters in many areas to reduce the costs of gathering needed weather information. The station consists of two major components. The first is the sensor group with meteorological sensors and data transmission electronics. The second is the converter display group that receives, displays, and records the data.

The sensor group consists of the signal conditioner, transmitter, and meteorological sensors including those for air temperature, dew point, and rainfall. The converter display group contains the data receiver and control assembly with a clock, a readout panel offering time, temperature, dew

point, maximum and minimum temperatures, and wind speed and direction. The analog recorder records these elements on a chart for future references (Fig. 6-14).

A variation of the automatic weather station is available to affluent amateur meteorologists. At least one major weather instrument supplier offers an automatic computerized weather station with several functions useful in forecasting weather at home (Fig. 6-15). The microprocessor-based weather station indicates time, indoor and outdoor temperatures, wind speed and direction, and barometric pressure on an LED display panel. It also offers average wind speed and automatically calculates wind chill factor. The unit's memory allows instant recall of date and time of maximum and minimum temperatures, of wind gusts, and of maximum and minimum barometric pressure. It even indicates the barometric pressure's rate of change per hour and tells if it is rising or falling. The unit is available in kit form for about $400 and fully assembled and tested for approximately $600, less cables.

RADAR

In the years following World War II, radar has attained increasing prominence in weather observations. The state-of-the-art of weather radar has grown from the experimental use of modified war surplus equipment to its current utilization as a valuable weather research, observing, and forecasting tool.

In 1947, the Weather Service began installing converted military radars, primarily in that section of the nation most frequently subject to tornado activity. In 1959, the WSR-57 (Weather Surveillance Radar-57) became available and a nationwide radar network was developed around this powerful, sensitive radar. They are manned 24 hours a day by observers trained to operate the sets, evaluate the echo displays, and disseminate weather information based on the displays. The system has been updated many times since.

Radar observations at network radar stations are transmitted at least hourly, while radar obser-

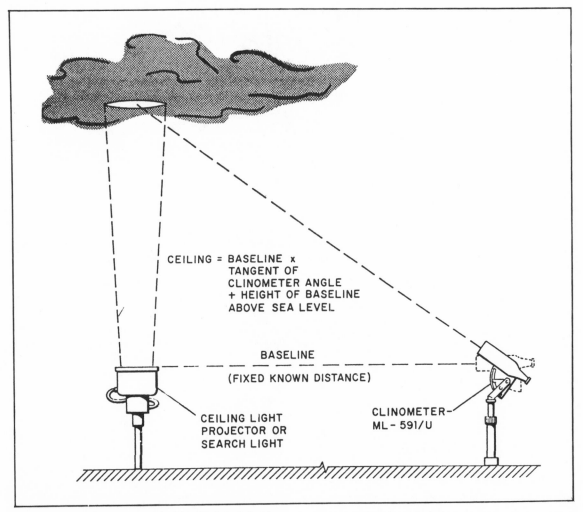

Fig. 6-11. Obtaining cloud base height with the clinometer.

116

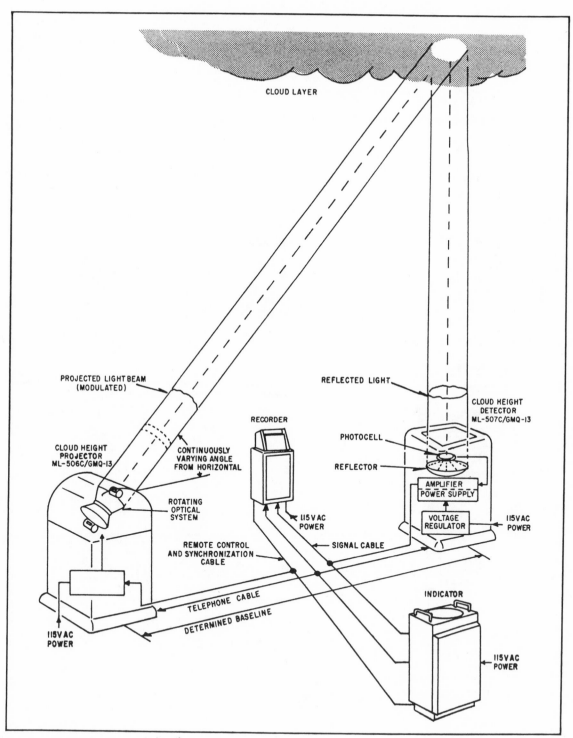

Fig. 6-12. Typical installation of the cloud height set.

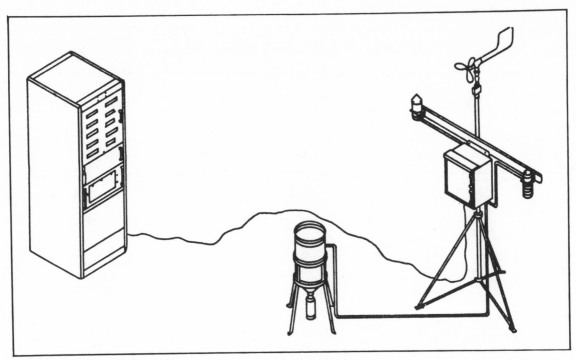

Fig. 6-13. Automatic weather station.

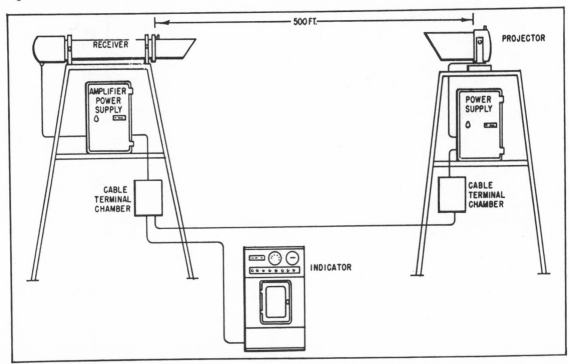

Fig. 6-14. Typical installation of a transmissometer.

Fig. 6-15. Digital weather computer. (courtesy Heath Company)

vations at local warning radar stations are taken on an as-needed basis. These observations consist of two parts: the polar coordinate section (used to brief pilots, by airlines and other Weather Service offices) and the digital section (used to produce the radar summary chart in Fig. 6-16).

The intensity and movement of severe local storms can be determined by observing the radar scope. The more intense the storm (as measured by reflectivity, height, and speed), the greater the chance of severe weather.

Most thunderstorms are made up of several cells; however, severe thunderstorms are usually single cells or super cells. Severe thunderstorms often have echo tops that penetrate the tropopause (over 50,000 feet). Storms of high reflectivity usually have tornadoes associated with them.

Precipitation associated with fronts does not always indicate the position of the front, and doesn't follow any set pattern. The widespread overrunning generally found with warm fronts often produces a large smooth echo. Radar is particularly useful in locating the thunderstorms occasionally imbedded in warm front weather. Cold front echoes are usually cellular in nature and are not always arranged in lines along the front. Radar is particularly useful in

following the movement of precipitation associated with cold fronts and in quickly detecting changes in precipitation patterns. Early detection of developing thunderstorms along a front or developing squall lines out ahead of the front can add important minutes or even hours to a warning service.

Hurricanes are the most spectacular of the large scale echo patterns. They are characterized by their well-known center, or *eye*, about which a great mass of moisture-laden air is swirling. When viewed on a radar scope, the eye of a mature hurricane is usually readily identified, appearing as a circular echo-free area. A well-developed storm is recognizable even when the center of the storm is a great distance from the radar, and when the center gets close enough that the wall cloud is detected, the hurricane is usually unmistakable. Also significant in hurricane identification are the outer rain bands, which often appear ahead of the large rain shield. Small tornadoes are sometimes found in these bands. Heavy rain can occur if the bands persist over a particular area. The rain shield is a potential flood threat, therefore it must be closely inspected for any change in size and location relative to the center. With an adequate radar network it is possible to have up-to-the-minute plots of the

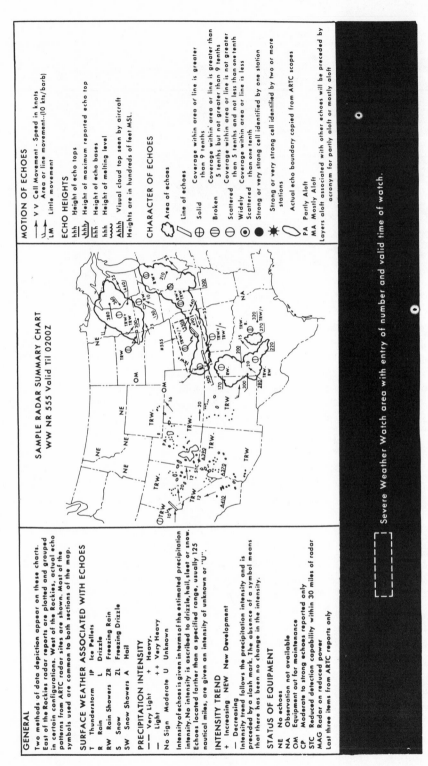

GENERAL

Two methods of data depiction appear on these charts. East of the Rockies radar reports are plotted and grouped in certain configurations. West of the Rockies, actual echo patterns from ARTC radar sites are shown. Most of the symbols used are common to both sections of the map.

SURFACE WEATHER ASSOCIATED WITH ECHOES

T	Thunderstorm	IP	Ice Pellets
R	Rain	L	Drizzle
RW	Rain Showers	ZR	Freezing Rain
S	Snow	ZL	Freezing Drizzle
SW	Snow Showers	A	Hail

PRECIPITATION INTENSITY

── Very Light   ─ Heavy.
─ Light   ++ Very Heavy
No Sign   Moderate   U   Unknown

Intensity of echoes is given in terms of the estimated precipitation intensity. No intensity is ascribed to drizzle, hail, sleet or snow. Echoes located farther than a specified range, usually 125 nautical miles, are given an intensity of unknown or "U".

INTENSITY TREND

+ Increasing   NEW   New Development
- Decreasing

Intensity trend follows the precipitation intensity and is preceded by a slash mark. The absence of a symbol means that there has been no change in the intensity.

STATUS OF EQUIPMENT

NE   No echoes
NA   Observation not available
OM   Equipment out for maintenance
CP   Moderate to strong echoes reported only
STC   Reduced detection capability within 30 miles of radar
MAG   Radar on reduced power
Last three items from ARTC reports only

SAMPLE RADAR SUMMARY CHART
WW NR 555 Valid Til 0200Z

MOTION OF ECHOES

↗ V V Cell Movement - Speed in knots
↗ Area or line movement--(10 kts/barb)
LM   Little movement

ECHO HEIGHTS

hhh   Height of echo tops
hhh   Height of maximum reported echo top
hhh   Height of echo bases
hhh   Height of melting level
Ahhh   Visual cloud top seen by aircraft
Heights are in hundreds of feet MSL.

CHARACTER OF ECHOES

Area of echoes
Line of echoes

Solid   Coverage within area or line is greater than 9 tenths
Broken   Coverage within area or line is greater than 5 tenths but not greater than 9 tenths
Scattered   Coverage within area or line is not greater than 5 tenths and not less than one tenth
Widely Scattered   Coverage within area or line is less than one tenth
Strong or very strong cell identified by one station
Strong or very strong cell identified by two or more stations
Actual echo boundary copied from ARTC scopes

PA   Partly Aloft
MA   Mostly Aloft
Layers aloft associated with other echoes will be preceded by acronym for partly aloft or mostly aloft

Severe Weather Watch area with entry of number and valid time of watch.

Fig. 6-16. Legend for radar summary chart.

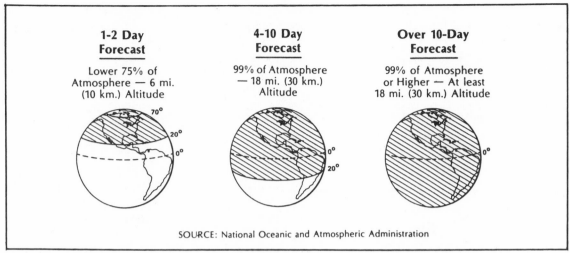

1-2 Day Forecast	4-10 Day Forecast	Over 10-Day Forecast
Lower 75% of Atmosphere — 6 mi. (10 km.) Altitude	99% of Atmosphere — 18 mi. (30 km.) Altitude	99% of Atmosphere or Higher — At least 18 mi. (30 km.) Altitude

SOURCE: National Oceanic and Atmospheric Administration

Fig. 6-17. Extent of information needed for a typical weather forecast.

hurricane center's position, and to detect any change in direction or speed of movement and intensity of the storm.

Squall lines, as you remember from Chapter 2, are non-frontal and resemble a violent cold front. They are usually found in the warm air ahead of the cold front and consist of a line of convective cells that can on occasion produce damaging wind, hail, or tornadoes. Squall lines exist because of unusual instability. Aircraft are particularly vulnerable to extreme instability, therefore they should plan to give all parts of a squall line a wide margin of safety.

There are currently three governmental departments using radar systems in the United States for related but not identical purposes: the Departments of Commerce, Defense, and Transportation. They are now attempting to merge their systems for increased efficiency and widened information. The new system is called Next Generation Weather Radar—acronymed NEXRAD—and is expected to offer more detailed and useful information on weather activity affecting daily transportation, commerce, and national defense (Fig. 6-17).

RADIOSONDE

Meteorologists need to know what the weather is like above 20,000 feet in order to accurately forecast tomorrow's weather conditions. The problem has been that, until recently, few planes and none of the early meteorograph kits could fly that high. Today, high altitude weather information is gathered with a system called *radiosonde* (with a silent "e").

The radiosonde (Fig. 6-18) consists of sensors used to measure the severel meteorological parameters coupled to a radio transmitter and assembled in a lightweight box. The sensing elements sample the ambient temperature, relative humidity, and pressure of the air through which it rises. The radio transmitter sends out a signal at 1680 MHz to stations on the ground so that the data can be evaluated instantaneously.

The radiosonde package is attached to a paper parachute and together they are suspended from a balloon. The balloon is inflated with a lighter-than-air gas (such as hydrogen, helium, or natural gas) and sent aloft. The balloon, parachute, and radiosonde (called a *flight train*) ascends at about 1000 feet per minute, continuously transmitting the temperature, relative humidity, and pressure to the ground-based Radiosonde Tracking System, which converts them into a continuous trace on the Strip Chart Recorder (Fig. 6-19). In this case, the curve A-B-C-D represents the temperature profile, and the curve 1-2-3-4 represents the relative humidity profile. The horizontal lines connecting them represent pressure.

Wind speed and direction are determined for

121

each minute of the flight, generally 90 minutes long. They are determined from changes in the position and direction of the flight train as detected by the tracking system. When winds are incorporated into the observation, it is called a *rawinsonde* observation.

When the balloon reaches its elastic limit—from 90,000 to 100,000 feet altitude—a small parachute slows the descent of the radiosonde, minimizing the danger to lives and property.

Significant changes in the meteorological

parameters are entered from the Strip Chart Recorder and the Data Printer into a minicomputer, which converts them into coded messages. These messages are relayed through various communication systems to all parts of the world. The information in these messages is extremely important in weather forecasting, climatology, and research. The National Meteorological Center (NMC) prepares daily upper air charts and forecasts based on the rawinsonde observations. After the data has been used it is archived at the National Climatic Center where it is available to anyone upon request.

Rawinsonde observations are taken twice daily at 94 National Weather Service stations in the United States, the Caribbean, and the Pacific Islands, and at 35 additional cooperative stations at various points in the Western Hemisphere. Figure 6-20 shows the network of National Weather Service and military rawinsonde stations in the 48 states.

So what happens to used radiosondes? About a third of them are found and returned to the Instrument Reconditioning Branch in Kansas City, Missouri, where they are repaired and reissued for further use—some as many as seven times. Instructions printed on the radiosonde explain the use of the instrument, state the approximate height reached and request the finder to mail the radiosonde—postage free—back to Uncle Sam. These recycled weather instruments reduce the overall cost of the system's operation.

WEATHER SATELLITES

The Meteorological Satellite is an unmanned space vehicle that orbits the earth carrying equipment to observe and report weather conditions on the earth below. The weather satellite is the greatest thing to happen to forecasting since the barometer (Fig. 6-21).

Weather satellites carry television cameras and infrared scanning radiometers to observe weather as it moves along the Earth. Signals from both systems are sent to tracking stations below. These stations translate the signals into pictures and maps that are used to forecast weather by "looking over the horizon."

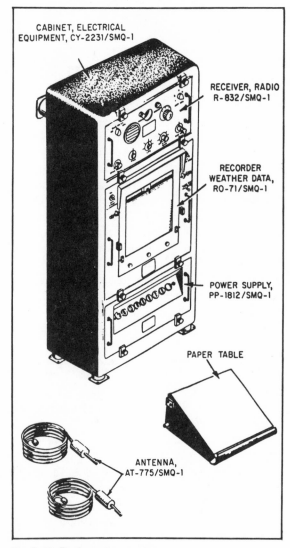

CABINET, ELECTRICAL EQUIPMENT, CY-2231/SMQ-1

RECEIVER, RADIO R-832/SMQ-1

RECORDER WEATHER DATA, RO-71/SMQ-1

POWER SUPPLY, PP-1812/SMQ-1

PAPER TABLE

ANTENNA, AT-775/SMQ-1

Fig. 6-18. Radiosonde receptor.

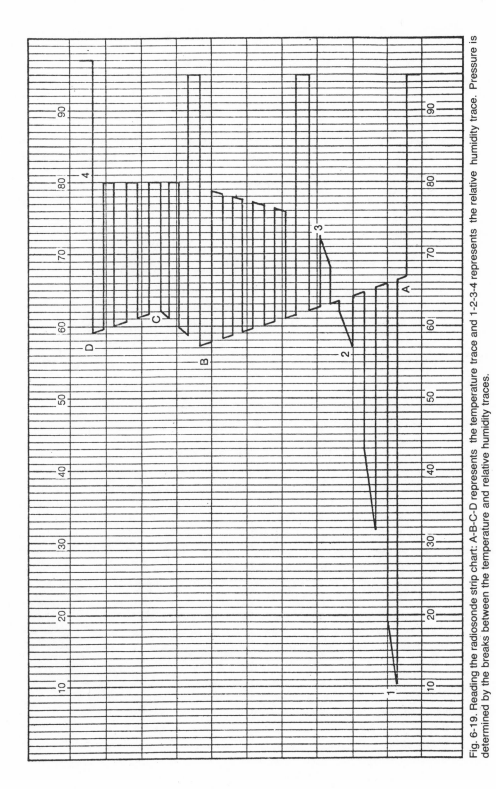

Fig. 6-19. Reading the radiosonde strip chart: A-B-C-D represents the temperature trace and 1-2-3-4 represents the relative humidity trace. Pressure is determined by the breaks between the temperature and relative humidity traces.

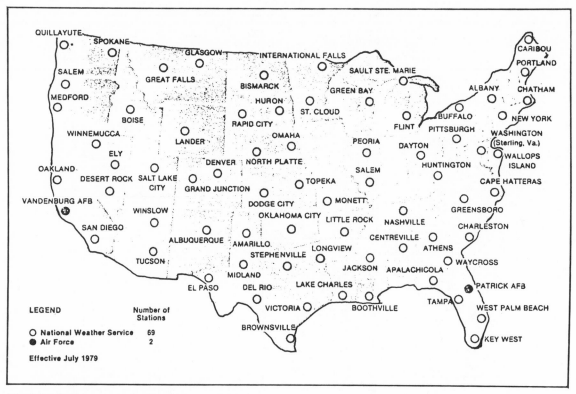

Fig. 6-20. Radiosonde stations in the United States.

The first satellite, Tiros I, is nearly a quarter-century old and circled the earth every 100 minutes at an altitude of about 450 miles (238,000 feet)—higher than any radiosonde balloon can go. You can see the pictures sent back by these satellites each evening on the weatherman's portion of the evening news. Many stations even offer time-lapse photographs showing how weather systems have moved into the area over the last few days (Fig. 6-22).

The United States provides satellite transmission directly to over three dozen countries through the World Meteorological Organization (WMO) to help other countries in using weather to their advantage.

WEATHER COMPUTERS

Computer technology has offered many blessings to the field of meteorology. An example is the Automation of Field Operations and Services (AFOS) program that is tested and now being spread by the National Weather Service. The AFOS is a computerized weather station that does away with teletypewriters, facsimile machines, and the enormous quantities of paper they required. Previ-

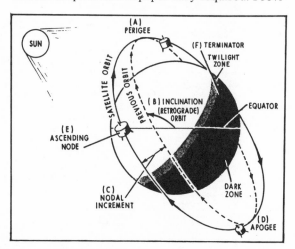

Fig. 6-21. Weather satellite terms defined.

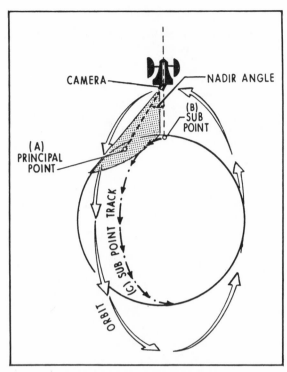

CAMERA — NADIR ANGLE

(B) SUB POINT

(A) PRINCIPAL POINT

(C) SUB POINT TRACK

ORBIT

Fig. 6-22. Weather satellite tracking terminology.

ously, forecasters spent vast amounts of time tearing off, sorting, and posting paper teletypewriter messages and paper maps. The AFOS system offers a weather map in 15 seconds rather than 10 minutes and sends messages at the rate of 3000 words per minute (compared to 100 words per minute). The message can go between the two most remote stations in about 25 seconds, with error checks at an average of 24 places between them.

With the new system, forecasters will have available a tremendous variety of weather maps and messages they can call up within seconds to aid them in preparing forecasts and warnings. The AFOS system is enhanced by other automated devices and systems such as automatic weather-observing stations, digitized radar, and computer-assisted radiosondes. These linkages allow fast and frequent observations of changes taking place in the weather.

NATIONAL METEOROLOGICAL CENTER

The Weather Service gathers data at hundreds

of locations on and over the earth's surface. At some point, this torrent of information must converge and be translated into a coherent picture of what the atmosphere is doing *now* and what it can be *expected* to do in the near future. The Weather Service's National Meteorological Center, or NMC, at Camp Springs, Maryland, is the necessary point of convergence. It is linked to the National Oceanic and Atmospheric Administration's largest electronic computers at Suitland, Maryland, a few miles away. There, millions of bits of information are collected, analyzed, and shaped into composite views of the world's weather—past, present, and future.

Operations at NMC are a near-unique blending of human skill and the special qualities of machines—high-speed computers, printers and electrostatic graphic devices and communications systems. Incoming observational data are processed on a large computer that identifies, checks, and sorts the various types of information and produces lists of error-free data in less than 10 minutes. A computer-driven printer produces a location chart, a discrepancy list of erroneous material, and an error-free list for checking, analysis, plotting, and filing.

As the stream of processed data moves on, it is sampled and interpreted by meteorologists and computers. Two cycles of machine analysis are run each day on observations taken over the Northern Hemisphere at 0000 and 1200 hours GMT (Greenwich Mean Time).

Each machine cycle begins with a preliminary analysis, made after an hour and a half of collection over North America from the surface to 30,000 feet. The operational analysis is made after three and a half hours of collection, when 80 percent of all data has come in. A final machine analysis is made twice a day after ten hours of collection, when most information is in, and covers the entire globe from the surface to 53,000 feet. A high-level analysis is done once a day for the region between 53,000 feet and 102,000 feet, for supersonic aviation.

Collection, preparation, and analysis of data are all preliminary to the task of forecasting what the atmosphere will be doing this afternoon, tomorrow, next week, or next month. NMC's meteorologists and machines next prepare twice-daily forecasts extended to 72 hours into the future, forecasts out to five days ahead daily, and out to a month ahead twice per month. They are also developing techniques for extending the forecast period and improving its reliability.

The products of NMC are the technical foundation for the local and regional forecasts issued by meteorologists in Weather Service offices and airport stations across the land. Rapid dissemination is an essential part of the product. To the weatherman in the field, this morning's prediction grows old as fast as today's newspapers.

The National Meteorological Center is also one of three designated World Meteorological Centers with global weather analysis and forecasting responsibilities.

As an American, these services are available to you 24 hours a day, offering weather observations that you cannot make yourself and helping you in forecasting and in taking advantage of weather conditions in your corner of the world.

Collecting weather information from all available sources is the topic of the next chapter.

Chapter 7

Collecting Weather Data

The first step in accurately forecasting weather is gathering all available meteorological information on the area you are considering. The sources for such information include the weather eye, and basic and advanced instruments discussed in Chapters 5 and 6. Before the data can be analyzed and interpreted it must be recorded and distributed.

THE WEATHER LOG

Don't trust your memory when making a weather observation; write down all observations. Try to take at least two readings of sky conditions and instruments each day. Additional readings will give an even better picture of changing weather patterns. Three-hourly or six-hourly observations are suitable for most amateur records.

Keep a weather log to record your observations (Fig. 7-1). It can be as simple or as detailed as you want to make it, depending on the observations to be made.

Here is a brief explanation of the entries to be made in a typical weather log using the basic instruments discussed in Chapter 5.

Sky: Enter the state of the sky (cloud cover) in tenths or substitute the following generally recognized weather symbols (Fig. 7-2):

0   Clear (less than 1/10).
◐   Scattered (2/10 to 5/10).
◑   Broken (5/10 to 9/10).
⊕   Overcast (more than 9/10).

If some form of precipitation is occurring, substitute one of the following letters for the sky symbol:

R   Rain.
S   Snow.
H   Hail.
T   Thunderstorm.
E   Sleet.
F   Fog.
Z   Freezing rain.
L   Drizzle.

Temperature: Read all thermometers to the

MONTH	SKY		TEMPERATURE					HUMIDITY		BAROMETER		WIND		Precipitation		Remarks
	AM	PM	AM	PM	Max.	Min.	Mean	AM	PM	AM	PM	AM	PM	AM	PM	
1																
2																
3																
4																

Fig. 7-1. Typical weather observation log.

nearest whole degree (C or F) and enter the current reading. Read the maximum thermometer at the p.m. observation and the minimum thermometer at the a.m. observation. Remember to reset these thermometers after each reading. Calculate the day's *mean* (average) temperature by adding the maximum and minimum temperatures and dividing by 2.

Humidity: Read wet and dry bulb thermometers of the psychrometer, consult the Table of Relative Humidity (Table 7-1), and enter as a whole percentage.

Comparison of the reading on the wet bulb thermometer with those on a dry bulb thermometer will show relative humidity by using this table. Top figures on the chart are the present dry bulb reading. By checking the left column for *degree difference* shown on the two thermometers and across to the dry bulb reading, the relative humidity can be found. For instance, if the difference between the dry and wet bulb readings is 6 degrees, and the dry bulb reading is 70 degrees, the relative humidity is 72 percent.

Barometer: Read the barometer to the nearest hundredth of an inch or half a millibar. If the barometer has been rising during the past three hours, mark a plus (+) after the reading entered; if falling, mark a minus (−) after the reading.

Wind: Observe the wind direction (Fig. 7-3) to eight points of the compass (N, NE, E, SE, S, SW, W, NW).

Read the anemometer or estimate the wind speed (Table 7-2). Enter the wind direction and speed in the appropriate column, i.e., NE-5 or SW-12.

Precipitation: Read the gauge and enter the amount of precipitation to the nearest tenth of an inch. If no precipitation has occurred, leave the column blank. Indicate a reading of less than a tenth of an inch of precipitation by a "T" for "trace." If precipitation is snow, take several readings on the ground with a yardstick. Make sure the measurements are not made in *drifted* snow. Average the readings and enter the number of inches followed by the symbol "*" to indicate snow. Generally speaking, ten inches of snow equals one inch of water or rain. Of course this varies, depending on the composition of the snow. Professional meteorologists often go through a more thorough process of melting the snow to learn its moisture content.

Be sure to empty the gauge after each observation.

Remarks: Enter any special phenomena such as lightning and thunder or smoke or haze and indicate the times that the event was observed. Entries can also be made here for the type of clouds in the sky and the direction of movement. A common system of noting clouds is:

CI	Cirrus.
CC	Cirrocumulus.
CS	Cirrostratus.
AC	Altocumulus.
AS	Altostratus.
NS	Nimbostratus.
SC	Stratocumulus.
ST	Stratus.
CU	Cumulus.
CB	Cumulonimbus.

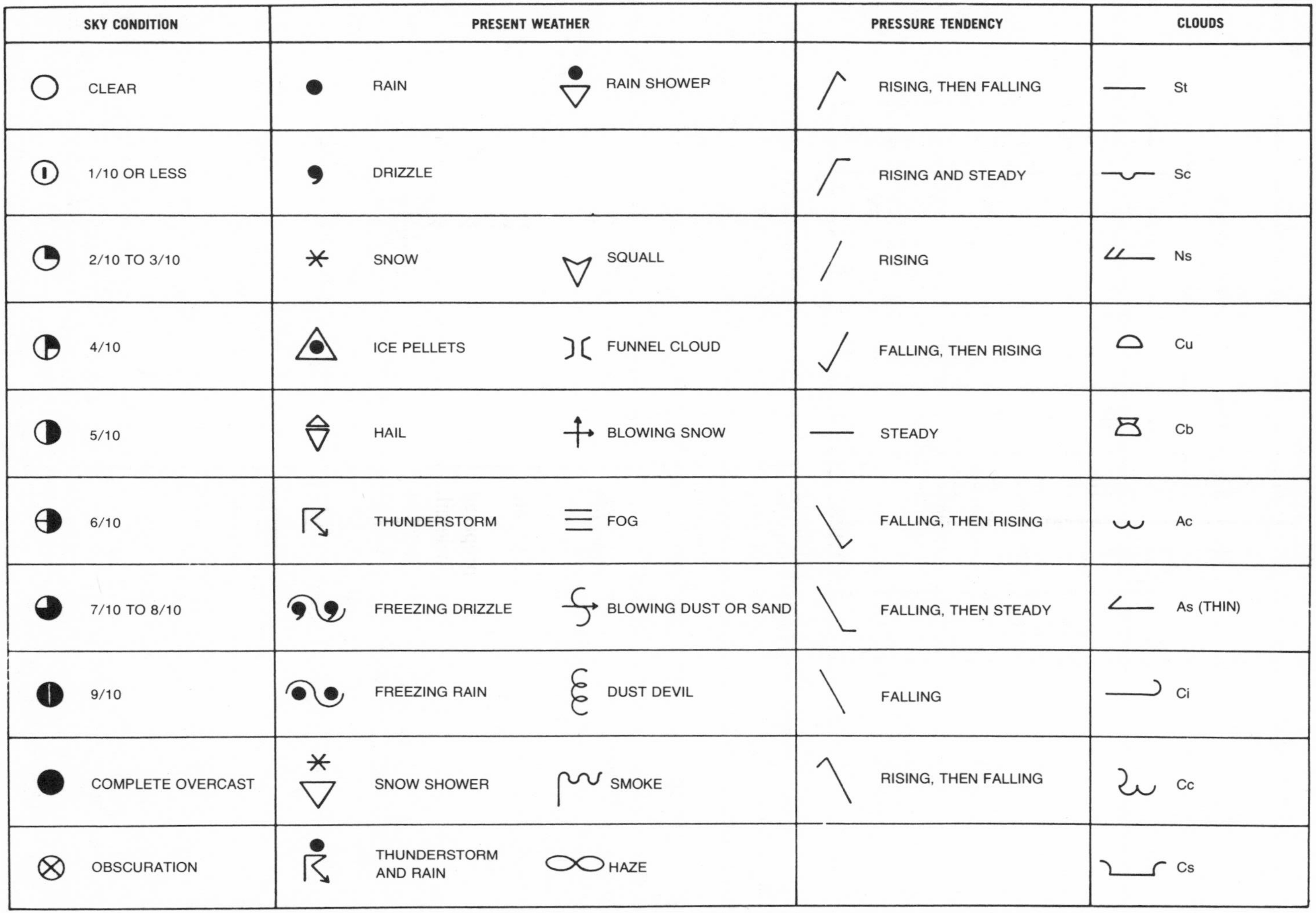

SKY CONDITION		PRESENT WEATHER		PRESSURE TENDENCY		CLOUDS	
CLEAR	○	RAIN ●	RAIN SHOWER ●▷	RISING, THEN FALLING	︿	St	—
1/10 OR LESS	⊖	DRIZZLE ●		RISING AND STEADY		Sc	
2/10 TO 3/10	◕	SNOW ✳	SQUALL ▽	RISING	/	Ns	⫽
4/10	◔	ICE PELLETS ◁	FUNNEL CLOUD)(	FALLING, THEN RISING	∨	Cu	⌐
5/10	◑	HAIL ◁	BLOWING SNOW ✛	STEADY	—	Cb	⬚
6/10	◕	THUNDERSTORM ⦦	FOG ≡	FALLING, THEN RISING	\	Ac	〕
7/10 TO 8/10	◕	FREEZING DRIZZLE ⦂	BLOWING DUST OR SAND ⬧	FALLING, THEN STEADY	\	As (THIN)	∨
9/10	◕	FREEZING RAIN ⦂	DUST DEVIL ⦡	FALLING	/	Ci	⌒
COMPLETE OVERCAST	●	SNOW SHOWER ✳▷	SMOKE ⌇	RISING, THEN FALLING	︿	Cc	⌇
OBSCURATION	⊗	THUNDERSTORM AND RAIN ●⦦	HAZE ∞			Cs	⎰

Fig. 7-2. The most common weather map symbols for sky condition, present weather, pressure tendency and clouds.

Table 7-1. Table of Relative Humidity.

Difference between wet-bulb and dry-bulb readings	Table of Relative Humidity Temperature of air, dry-bulb thermometer, Fahrenheit							
	30°	40°	50°	60°	70°	80°	90°	100°
1	90	92	93	94	95	96	96	97
2	79	84	87	89	90	92	92	93
3	68	76	80	84	86	87	88	90
4	58	68	74	78	81	83	85	86
6	38	52	61	68	72	75	78	80
8	18	37	49	58	64	68	71	71
10		22	37	48	55	61	65	68
12		8	26	39	48	54	59	62
14			16	30	40	47	53	57
16			5	21	33	41	47	51
18				13	26	35	41	47
20				5	19	29	36	42
22					12	23	32	37
24					6	18	26	33

Or, you may want to use the abbreviations and International Code signs used by the National Weather Service offered later.

COLLECTING DATA FROM THE WEATHER SERVICE

The amateur meteorologist can supplement observed and recorded weather with information from the world's largest system of weather data collection, the National Weather Service. Forecasts, outlooks, and raw data are available from the Weather Service in a number of ways.

Forecasts

Forecasts are detailed predictions of expected weather conditions for periods of time up to five days in advance. They may refer to an individual city (Chicago forecast) or an area as large as a state (Vermont forecast).

Local, zone, and state forecasts are prepared several times a day to predict the weather, in detail, for the next 36 to 48 hours. A "local" forecast applies to a city and its immediate surroundings. A "zone" forecast applies to a small portion of a state. A "state" forecast applies to an entire state (Table 7-3).

Extended forecasts are issued daily to predict weather over a state in board terms for a period three days beyond that of the state forecast.

Travelers' forecasts are prepared in two forms. One is issued nationwide for about 90 cities and consists of a one-word statement on the expected weather and the forecast high and low temperature for the next 36 to 48 hours for each city.

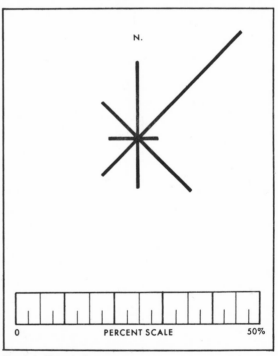

Fig. 7-3. The wind rose.

130

The other is issued by some local National Weather Service offices to describe general driving conditions over routes within a few hundred miles of the city where the office is located.

Recreation forecasts are issued by selected NWS offices based on local needs and office capabilities. Forecasts for boating and skiing areas are two of the more popular examples.

Weather watches and warnings are issued in connection with hazardous weather—tornadoes, winter storms, hurricanes, flash floods, and the like. A *watch* is issued, where possible, well in advance to alert the public to the *possibility* of dangerous weather developing. A *warning* is normally issued when the particular hazard is *imminent* or *reported*.

River and flood forecasts and warnings provide daily information on stages of major rivers and streams and, when necessary, warnings of potential flood conditions.

Outlooks

Outlooks are estimates of general weather conditions for periods beyond five days from the date of issue. Naturally, such predictions have less precision and detail than short-term forecasts. They are useful for general planning, however.

Six- to ten-day outlooks are issued three times a week. They give estimates of temperature and

Table 7-2. Wind Speed Table.

Wind speed		Seaman's term	World meteoro-logical organi-zation (1964)	Estimating wind speed		World meteorological organization		
Knots	Km per hour			Effects observed at sea	Effects observed on land	Term	Code	Height of waves in feet
under 1	under 1	Calm	Calm	Sea like mirror.	Calm; smoke rises vertically.	Calm, glassy	0	0
1-3	1-5	Light air	Light air	Ripples with appearance of scales; no foam crests.	Smoke drift indicates wind direction; vanes do not move.			
4-6	6-11	Light breeze	Light breeze	Small wavelets; crests of glassy appearance, not breaking.	Wind felt on face; leaves rustle; vanes begin to move.	Calm, rippled	1	0-1/3
7-10	12-19	Gentle breeze	Gentle breeze	Large wavelets; crests begin to break; scattered whitecaps.	Leaves, small twigs in constant motion; light flags extended.	Smooth, wavelets	2	1/3-1 2/3
11-16	20-28	Moderate breeze	Moderate breeze	Small waves, becoming longer; numerous whitecaps.	Dust, leaves, and loose paper raised up; small branches move.	Slight	3	2-4
17-21	29-38	Fresh breeze	Fresh breeze	Moderate waves, taking longer form; many whitecaps; some spray.	Small trees in leaf begin to sway.	Moderate	4	4-8
22-27	39-49	Strong breeze	Strong breeze	Larger waves forming; whitecaps everywhere; more spray.	Larger branches of trees in motion; whistling heard in wires.	Rough	5	8-13
28-33	50-61	Moderate gale	Near gale	Sea heaps up; white foam from breaking waves begins to be blown in streaks.	Whole trees in motion; resistance felt in walking against wind.	Very rough	6	13-20
34-40	62-74	Fresh gale	Gale	Moderately high waves of greater length; edges of crests begin to break into spindrift; foam is blown in well-marked streaks.	Twigs and small branches broken off trees; progress generally impeded.			
41-47	75-88	Strong gale	Strong gale	High waves; sea begins to roll; dense streaks of foam; spray may reduce visibility.	Slight structural damage occurs; slate blown from roofs.			
48-55	89-102	Whole gale	Storm	Very high waves with overhanging crests; sea takes white appearance as foam is blown in very dense streaks; rolling is heavy and visibility reduced.	Seldom experienced on land; trees broken or uprooted; considerable structural damage occurs.	High	7	20-30
56-63	103-117	Storm	Violent storm	Exceptionally high waves; sea covered with white foam patches; visibility still more reduced.	Very rarely experienced on land; usually accompanied by widespread damage.	Very high	8	30-45
64-71 72-80 81-89 90-99 100-108 109-118	118-133 134-149 150-166 167-183 184-201 202-220	Hurricane	Hurricane	Air filled with foam; sea completely white with driving spray; visibility greatly reduced.		Phenomenal	9	over 45

Table 7-3. Time Conversion Table.

Previous day	1800	1900	2000	2100	2200	2300	2400	0100	0200	0300	0400	0500	0600	0700	0800	0900	1000	1100	1200	1300	1400	1500	1600	1700	1800	**Same day**
	1900	2000	2100	2200	2300	2400	0100	0200	0300	0400	0500	0600	0700	0800	0900	1000	1100	1200	1300	1400	1500	1600	1700	1800	1900	
	2000	2100	2200	2300	2400	0100	0200	0300	0400	0500	0600	0700	0800	0900	1000	1100	1200	1300	1400	1500	1600	1700	1800	1900	2000	
	2100	2200	2300	2400	0100	0200	0300	0400	0500	0600	0700	0800	0900	1000	1100	1200	1300	1400	1500	1600	1700	1800	1900	2000	2100	
	2200	2300	2400	0100	0200	0300	0400	0500	0600	0700	0800	0900	1000	1100	1200	1300	1400	1500	1600	1700	1800	1900	2000	2100	2200	
	2300	2400	0100	0200	0300	0400	0500	0600	0700	0800	0900	1000	1100	1200	1300	1400	1500	1600	1700	1800	1900	2000	2100	2200	2300	
	2400	0100	0200	0300	0400	0500	0600	0700	0800	0900	1000	1100	1200	1300	1400	1500	1600	1700	1800	1900	2000	2100	2200	2300	2400	
Same day	0100	0200	0300	0400	0500	0600	0700	0800	0900	1000	1100	1200	1300	1400	1500	1600	1700	1800	1900	2000	2100	2200	2300	2400	0100	**Next day**
	0200	0300	0400	0500	0600	0700	0800	0900	1000	1100	1200	1300	1400	1500	1600	1700	1800	1900	2000	2100	2200	2300	2400	0100	0200	
	0300	0400	0500	0600	0700	0800	0900	1000	1100	1200	1300	1400	1500	1600	1700	1800	1900	2000	2100	2200	2300	2400	0100	0200	0300	
	0400	0500	0600	0700	0800	0900	1000	1100	1200	1300	1400	1500	1600	1700	1800	1900	2000	2100	2200	2300	2400	0100	0200	0300	0400	
	0500	0600	0700	0800	0900	1000	1100	1200	1300	1400	1500	1600	1700	1800	1900	2000	2100	2200	2300	2400	0100	0200	0300	0400	0500	
	0600	0700	0800	0900	1000	1100	1200	1300	1400	1500	1600	1700	1800	1900	2000	2100	2200	2300	2400	0100	0200	0300	0400	0500	0600	
	0700	0800	0900	1000	1100	1200	1300	1400	1500	1600	1700	1800	1900	2000	2100	2200	2300	2400	0100	0200	0300	0400	0500	0600	0700	
	0800	0900	1000	1100	1200	1300	1400	1500	1600	1700	1800	1900	2000	2100	2200	2300	2400	0100	0200	0300	0400	0500	0600	0700	0800	
	0900	1000	1100	1200	1300	1400	1500	1600	1700	1800	1900	2000	2100	2200	2300	2400	0100	0200	0300	0400	0500	0600	0700	0800	0900	
	1000	1100	1200	1300	1400	1500	1600	1700	1800	1900	2000	2100	2200	2300	2400	0100	0200	0300	0400	0500	0600	0700	0800	0900	1000	
	1100	1200	1300	1400	1500	1600	1700	1800	1900	2000	2100	2200	2300	2400	0100	0200	0300	0400	0500	0600	0700	0800	0900	1000	1100	
	1200	1300	1400	1500	1600	1700	1800	1900	2000	2100	2200	2300	2400	0100	0200	0300	0400	0500	0600	0700	0800	0900	1000	1100	1200	
	1300	1400	1500	1600	1700	1800	1900	2000	2100	2200	2300	2400	0100	0200	0300	0400	0500	0600	0700	0800	0900	1000	1100	1200	1300	
	1400	1500	1600	1700	1800	1900	2000	2100	2200	2300	2400	0100	0200	0300	0400	0500	0600	0700	0800	0900	1000	1100	1200	1300	1400	
	1500	1600	1700	1800	1900	2000	2100	2200	2300	2400	0100	0200	0300	0400	0500	0600	0700	0800	0900	1000	1100	1200	1300	1400	1500	
	1600	1700	1800	1900	2000	2100	2200	2300	2400	0100	0200	0300	0400	0500	0600	0700	0800	0900	1000	1100	1200	1300	1400	1500	1600	
	1700	1800	1900	2000	2100	2200	2300	2400	0100	0200	0300	0400	0500	0600	0700	0800	0900	1000	1100	1200	1300	1400	1500	1600	1700	
	Y	X	W	V	U	T	S	R	Q	P	O	N	Z	A	B	C	D	E	F	G	H	I	K	L	M	
	+12	+11	+10	+9	+8	+7	+6	+5	+4	+3	+2	+1	0	−1	−2	−3	−4	−5	−6	−7	−8	−9	−10	−11	−12	

precipitation for each state in very broad terms (with respect to normal).

Average monthly weather outlooks are issued shortly before the 1st and 15th of each month. They deal with changes from the average in temperature and precipitation over broad areas of the nation for the next 30 days.

Seasonal outlooks are issued during the last week of February, May, August, and November, projecting temperature changes from the average for the following three months. They occasionally include a precipitation outlook.

Raw Weather Data

National Weather Service offices can also be your auxiliary source of weather data. Through various distribution methods to be discussed in a moment, the NWS offers current information on precipitation, temperature, pressure, wind direction and speed, humidity, dew point, clouds, visibility, ceiling, and other valuable data. In fact, by using weather information distributed free or at minimal cost, you can forecast local and even regional weather accurately without a single weather observation instrument.

DISTRIBUTION OF WEATHER INFORMATION

National Weather Service products are dis-

tributed in a number of ways. In addition to the systems operated by local weather offices, information is given to the news media and the telephone industry for relay to the public.

The primary outlet for NWS forecasts and warnings is radio, television, and newspapers. Coverage varies greatly from station to station and paper to paper because dissemination of weather information is a voluntary service. Many cable TV systems have a channel devoted entirely or in part to weather. Figure 7-4 shows a typical newspaper's weather column, offering area weather conditions and forecasts, regional and national weather conditions, river and tide conditions, and a simplified national weather map.

Printed weather information is also available directly from the National Weather Service. The *Average Monthly Weather Outlook* (including the Seasonal Outlook), *Daily Weather Maps, Weekly Series,* and other publications are available through the Superintendent of Documents, U.S. Government Printing Office, Washington, D.C., 20402, on a fee basis. Subject Bibliography (SB) 234, entitled *Weather,* is available from the same source. It offers a list and prices of meteorological publications available through the government.

Telephone recordings are available at many National Weather Service offices. For a listing of

WEATHER

In our area

TEMPERATURE   High   Low
SW Research Center, Hazel Dell
In 24 hours to 8 a.m.   96   53
Weather Service, Portland
In 24 hours to 8 a.m.   99   59
Yesterday's record   95   48
                                   1958   1975

PRECIPITATION   Inches
SW Research Center, Hazel Dell
In 24 hours to 8 a.m.   none
Cumulative this year   24.45
To this date last year   20.12
Weather Service, Portland
In 24 hours to 8 a.m.   none

River level

The Columbia River reached 13.1 feet below flood stage at 7 a.m. today at the Interstate 5 Bridge.

Local skies

WEDNESDAY, AUGUST 25

Sunset today . . . . 8:03 p.m.
Sunrise tomorrow . . . . 6:24 a.m.
Moonset tonight . . . . 11:40 p.m.
First quarter tomorrow . . . . . 2:49 a.m.

The planet Mercury is low in the west at sunset and sets in the early evening twilight. Do not be disappointed if you cannot see the elusive planet; it is making about its poorest appearance of the year.

Forecasts

Vancouver — Portland — Fair and cooler today with highs in the upper 80s. Fair Thursday after morning clouds. Highs near 80. Low 60. Southwest to west winds 10-20 mph.

Vancouver north to Olympia — Fair tonight except for patchy fog or low clouds, lows in the 50s and winds variable 5-15 mph. Fair Thursday after morning fog or low clouds, highs in the 70s.

Willamette Valley — Partly cloudy tonight. Lows 55 to 60. Variable cloudiness. Highs about 80.

Columbia Gorge — Variable cloudiness through Thursday. Lows near 60. Highs low 80s west to low 90s east. West wind to 20 mph.

Regional

24 hours to 4 a.m. today

	High	Low	Pr
Astoria	82	58	
Bellingham	80	56	
Colville	89	57	
Eugene	96	54	
Hoquiam	90	58	
Lakeview	92	53	
Medford	101	59	
Newport	62	52	
North Bend	64	57	
Olympia	94	52	
Portland	99	63	
Redmond	90	53	
Salem	97	54	
Seattle	88	62	
Spokane	90	55	
Walla Walla	96	64	
Wenatchee	93	62	
Yakima	90	51	

National

24 hours to 4 a.m. today PDT

	High	Low	Prc	Otlk
Albany	82	58	.01	clr
Albuquerque	93	64	.16	cdy
Amarillo	88	59	.20	cdy
Anchorage	61	49	.07	cdy

The Forecast For 8 p.m. EDT
Thursday, August 26
● High Temperatures

Rain · Snow · Showers · Flurries

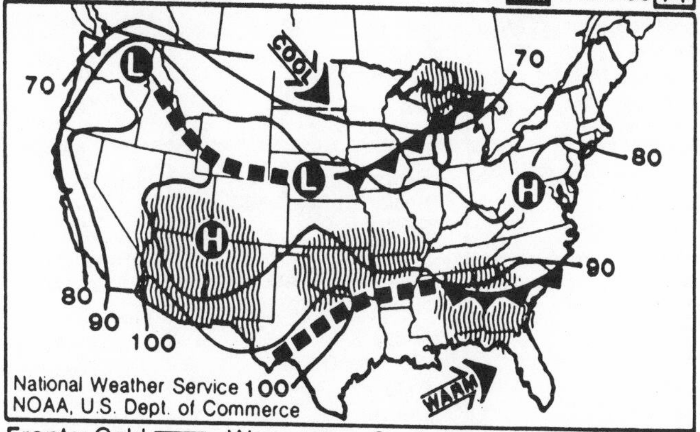

National Weather Service
NOAA, U.S. Dept. of Commerce

Fronts: Cold · Warm · Occluded · Stationary

	High	Low	Prc	Otlk
Asheville	84	67	.51	cdy
Atlanta	92	73		cdy
Atlantic City	80	68	.51	clr
Austin	100	73		clr
Baltimore	87	71	.12	clr
Billings	82	60		cdy
Birmingham	91	72		clr
Bismarck	74	49		cdy
Boise	66	55		cdy
Boston	85	66		clr
Brownsville	96	79		cdy
Buffalo	76	62	1.46	cdy
Burlington	77	61		clr
Casper	78	50		cdy
Charleston, S.C.	94	80		cdy
Charleston, W.Va.	85	72	.07	clr
Charlotte, N.C.	93	75	.26	cdy
Cheyenne	69	49		clr
Chicago	75	54	1.06	cdy
Cincinnati	82	68	.17	clr
Cleveland	84	65	.13	clr
Columbia, S.C.	94	75		cdy
Columbus	83	69	.16	clr
Dallas-Fort Worth	101	76		clr
Dayton	81	65	1.54	clr
Denver	74	55		cdy
Des Moines	79	55	.03	cdy
Detroit	80	61	.01	cdy
Duluth	72	43	.09	cdy
El Paso	96	71	.05	cdy
Fairbanks	73	49		cdy
Fargo	74	49		cdy
Flagstaff	77	56	.39	rn
Great Falls	78	56	1.26	cdy
Hartford	84	62	.02	clr
Helena	86	48		cdy
Honolulu	89	75		clr
Houston	94	78		clr
Indianapolis	78	63	.25	cdy
Jackson, Miss.	95	74		clr
Jacksonville	95	75	.15	cdy
Juneau	58	51	.20	rn
Kansas City	79	56	.21	cdy
Knoxville	86	77	.10	cdy
Las Vegas	80	68	.22	cdy
Little Rock	96	75		cdy
Los Angeles	75	66		cdy
Louisville	86	68	.04	cdy
Lubbock	97	65		cdy
Memphis	94	80		rn
Miami	88	83		cdy
Milwaukee	70	55	.35	cdy
Minneapolis-St. Paul	76	53	.56	cdy
Nashville	90	71	.70	rn
New Orleans	95	73		cdy
New York	84	70	.30	clr
Norfolk	85	74		clr

	High	Low	Prc	Otlk
North Platte	75	47		cdy
Oklahoma City	102	69		cdy
Omaha	74	54	.18	cdy
Orlando	93	76		cdy
Philadelphia	85	70	.19	clr
Phoenix	96	75	.65	cdy
Pittsburgh	81	68	.52	clr
Portland, Maine	80	60	.01	clr
Providence	83	63		clr
Raleigh	91	75		cdy
Rapid City	73	50		cdy
Reno	97	54		clr
Richmond	87	73		clr
Salt Lake City	89	65		clr
San Antonio	100	76		clr
San Diego	74	71		cdy
San Francisco	60	54		clr
Shreveport	96	74		cdy
Sioux Falls	76	53		cdy
St. Louis	82	59	.05	cdy
St. Pete-Tampa Bay	91	78		cdy
Sault Ste. Marie	63	42		cdy
Syracuse	80	61	.35	cdy
Topeka	80	53	.09	cdy
Tucson	88	66	.35	cdy
Tulsa	104	69		cdy
Washington	87	74		clr
Wichita	86	62		cdy

Extremes, excluding Alaska:
High: 105   Fort Sill, Oklahoma
Low: 30   West Yellowstone, Montana

Precipitation — Pr.   Rain — rn
Outlook — Otlk.   Snow — sn
Cloudy — cdy   Clear — clr

Tides

High Tides at Astoria

Aug. 26	6:59 a.m.,	5.5	6:55 p.m.,	7.2
Aug. 27	8:09 a.m.,	5.3	7:53 p.m.,	7.1
Aug. 28	9:21 a.m.,	5.4	8:52 p.m.,	7.1
Aug. 29	10:23 a.m.,	5.6	9:51 p.m.,	7.2
Aug. 30	11:12 a.m.,	5.9	10:43 p.m.,	7.4
Aug. 31	11:58 a.m.,	6.3	11:30 p.m.,	7.6
Sept 1			12:36 p.m.,	6.6

Low Tides at Astoria

Aug. 26	0:54 a.m.,	0.7	12:39 p.m.,	2.2
Aug. 27	1:53 a.m.,	0.6	1:38 p.m.,	2.7
Aug. 28	2:55 a.m.,	0.5	2:45 p.m.,	2.9
Aug. 29	3:56 a.m.,	0.2	3:51 p.m.,	2.9
Aug. 30	4:51 a.m.,	-0.1	4:48 p.m.,	2.7
Aug. 31	5:38 a.m.,	-0.4	5:41 p.m.,	2.4
Sept 1	6:19 a.m.,	-0.6	6:25 p.m.,	2.0

Deduct these hours and minutes from tables:
Long Beach, 1:10; Clatsop beaches, 0:50; Tillamook, 0:45; Yaquina, 1:00; Newport, 1:15; Vancouver add 5:45 for high, 7:24 for low.

Fig. 7-4. Typical newspaper weather map and information.

Table 7-4. NOAA National Weather Radio Network Station Locations and Frequencies.

Legend—Frequencies are identified as follows:
(1)—162.550 MHz
(2)—162.400 MHz
(3)—162.475 MHz
(4)—162.425 MHz
(5)—162.450 MHz
(6)—162.500 MHz
(7)—162.525 MHz

Location	Frequency
Alabama	
Anniston	3
Birmingham	1
*Columbia	4
Demopolis	3
Dozier	1
Florence	3
Huntsville	2
Louisville	3
Mobile	1
Montgomery	2
Tuscaloosa	2
Alaska	
Anchorage	1
Cordova	1
Fairbanks	1
Homer	2
Juneau	1
Ketchikan	1
Kodiak	1
Nome	1
Petersburg	1
Seward	1
Sitka	2
Valdez	1
Wrangell	2
Yakutat	1
Arizona	
Flagstaff	2
Phoenix	1
Tucson	2
Yuma	1
Arkansas	
Mountain View	2
Fayetteville	3
Fort Smith	2
Gurdon	3
Jonesboro	1
Little Rock	1
Star City	2
Texarkana	1
California	
Bakersfield (P)	1
Coachella (P)	2

Eureka	2
Fresno	2
Los Angeles	1
Merced	1
Monterey	2
Point Arena	2
Redding (P)	1
Sacramento	2
San Diego	2
San Francisco	1
San Luis Obispo	1
Santa Barbara	2
Colorado	
Alamosa (P)	3
Colorado Springs	3
Denver	1
Grand Junction	1
Greeley	2
Longmont	1
Pueblo	2
Sterling	2
Connecticut	
Hartford	3
Meriden	2
New London	1
Delaware	
Lewes	1
District of Columbia	
Washington, D.C.	1
Florida	
Clewiston	2
Daytona Beach	2
Fort Myers	3
Gainesville	3
Jacksonville	1
Key West	2
Melbourne	1
Miami	1
Orlando	3
Panama City	1
Pensacola	2
Tallahassee	2
Tampa	1
West Palm Beach	3
Georgia	
Athens	2
Atlanta	1
Augusta	1
*Baxley	7
Chatsworth	2
Columbus	2
Macon	3
Pelham	1
Savannah	2
Waycross	3

Hawaii	
Hilo	1
Honolulu	1
Kokee	2
Mt. Haleakala	2
Waimanalo (R)	2
Idaho	
Boise	1
Lewiston (P)	1
Pocatello	1
Twin Falls	2
Illinois	
Champaign	1
Chicago	1
Marion	4
Moline	1
Peoria	3
Rockford	3
Springfield	2
Indiana	
Evansville	1
Fort Wayne	1
Indianapolis	1
Lafayette	3
South Bend	2
Terre Haute	2
Iowa	
Cedar Rapids	3
Des Moines	1
Dubuque (P)	2
Sioux City	3
Waterloo	1
Kansas	
Chanute	2
Colby	3
Concordia	1
Dodge City	3
Ellsworth	2
Topeka	3
Wichita	1
Kentucky	
Ashland	1
Bowling Green	2
Covington	1
Elizabethtown (R)	2
Hazard	3
Lexington	2
Louisville	3
Mayfield	3
Pikeville (R)	2
Somerset	1
Louisiana	
Alexandria	3
Baton Rouge	2
Buras	3

Lafayette	1
Lake Charles	2
Monroe	1
Morgan City	3
New Orleans	1
Shreveport	2
Maine	
*Dresden	3
Ellsworth	2
Portland	1
Maryland	
Baltimore	2
Hagerstown	3
Salisbury	3
Massachusetts	
Boston	3
Hyannis	1
Worcester	1
Michigan	
Alpena	1
Detroit	1
Flint	2
Grand Rapids	1
Houghton	2
Marquette	1
Onondaga	2
Sault Sainte Marie	1
Traverse City	2
Minnesota	
Detroit Lakes	3
Duluth	1
International Falls	1
Mankato	2
Minneapolis	1
Rochester	3
Saint Cloud (P)	3
Thief River Falls	1
Willmar (P)	2
Mississippi	
Ackerman	3
Booneville	1
Bude	1
Columbia (R)	2
Gulfport	2
Hattiesburg	3
Inverness	1
Jackson	2
Meridian	1
Oxford	2
Missouri	
Columbia	2
Camdenton	1
Hannibal	3
Joplin/Carthage	1
Kansas City	1

St. Joseph	2
St. Louis	1
Sikeston	2
Springfield	2

Montana

Billings	1
Butte	1
Glasgow	1
Great Falls	1
Havre (P)	2
Helena	2
Kalispell	1
Miles City	2
Missoula	2

Nebraska

Bassett	3
Grand Island	2
Holdrege	3
Lincoln	3
Merriman	2
Norfolk	1
North Platte	1
Omaha	2
Scottsbluff	1

Nevada

Elko	1
Ely	2
Las Vegas	1
Reno	1
Winnemucca	2

New Hampshire

Concord	2

New Jersey

Atlantic City	2

New Mexico

Albuquerque	2
Clovis	3
Des Moines	1
Farmington	3
Hobbs	2
Las Cruces	2
Ruidoso	1
Santa Fe	1

New York

Albany	1
Binghamton	3
Buffalo	1
Elmira	1
Kingston	3
New York City	1
Rochester	2
Syracuse	1

North Carolina

Asheville	2
Cape Hatteras	3
Charlotte	3
Fayetteville	3
New Bern	2
Raleigh/Durham	1
Rocky Mount	3
Wilmington	1
Winston-Salem	2

North Dakota

Bismarck	2
Dickinson	2
Fargo	2
Jamestown	2
Minot	2
Petersburg	2
Williston	2

Ohio

Akron	2
Caldwell	3
Cleveland	1
Columbus	1
Dayton	3
Lima	2
Sandusky	2
Toledo	1

Oklahoma

Clinton	3
Enid	3
Lawton	1
McAlester	3
Oklahoma City	2
Tulsa	1

Oregon

Astoria	2
Brookings	1
Coos Bay	2
Eugene	2
Klamath Falls	1
Medford	2
Newport	1
Pendleton	1
Portland	1
Roseburg	3
Salem	3

Pennsylvania

Allentown	2
Clearfield	1
Erie	2
Harrisburg	1
Johnstown	2
Philadelphia	3
Pittsburgh	1
State College	3
Wilkes-Barre	1
Williamsport	2

Puerto Rico

Maricao	1
San Juan	2

Rhode Island

Providence	2

South Carolina

Beaufort	3
Charleston	1
Columbia	2
Florence	1
Greenville	1
Myrtle Beach	2
Sumter (R)	3

South Dakota

Aberdeen	3
Huron	1
Pierre	2
Rapid City	1
Sioux Falls	2

Tennessee

Bristol	1
Chattanooga	1
Cookeville	2
Jackson	1
Knoxville	3
Memphis	3
Nashville	1
Shelbyville	3
Waverly	2

Texas

Abilene	2
Amarillo	1
Austin	2
Beaumont	3
Big Spring	3
Brownsville	1
Bryan	1
Corpus Christi	1
Dallas	2
Del Rio (P)	2
El Paso	3
Fort Worth	1
Galveston	1
Houston	2
Laredo	3
Lubbock	2
Lufkin	1
Midland	2
Paris	1
Pharr	2
San Angelo	1
San Antonio	1
Sherman	3
Tyler	3
Victoria	2
Waco	3
Wichita Falls	3

Utah

Logan	2
Cedar City	2
Vernal	2
Salt Lake City	1

Vermont

Burlington	2
Windsor	3

Virginia

Heathsville	2
Lynchburg	2
Norfolk	1
Richmond	3
Roanoke	3

Washington

Neah Bay	1
Olympia	3
Seattle	1
Spokane	2
Wenatchee	3
Yakima	1

West Virginia

(see note 5)	
Charleston	2
Clarksburg	1

Wisconsin

La Crosse (P)	1
Green Bay	1
Madison	1
Menomonie	2
Milwaukee	2
Wausau	3

Wyoming

Casper	1
Cheyenne	3
Lander	3
Sheridan (P)	3

Notes:

1. Stations marked with an asterisk (*) are funded by public utility companies.

2. Stations marked (R) are low powered experimental repeater stations serving a very limited local area.

3. Stations marked (P) operate less than 24 hours/day; however, hours are extended when possible during severe weather.

4. Occasionally the frequency of an existing or planned station must be changed because of unexpected radio frequency interference with adjacent NOAA Weather Radio stations and/or with other government or commercial operators within the area.

5. Six additional stations are planned for West Virginia: Gilbert, Beckley, Hinton, Spencer, Sutton and Romney. Frequencies had not been assigned at the time of this printing.

these numbers see your city's phone directory under "United States Government, Department of Commerce, National Weather Service," or "National Weather Service," or simply "Weather." Large capacity weather information recordings are operated at some locations by telephone companies with forecasts supplied by the National Weather Service. For these listings, see the city's directory under "Weather" or front pages of the directory.

The long-distance information operator can provide you with the numbers of weather recordings in other cities, where available.

Weather Radio

The National Oceanic and Atmospheric Administration (NOAA)—parent of the National Weather Service—offers continuous weather information directly from about 350 NWS offices. Taped weather messages are repeated every four to six minutes and are routinely revised every one to three hours, or more frequently if needed. Most of the stations operate 24 hours daily.

The broadcasts are tailored to the weather information needs of people within the receiving area. For example, stations along the sea coasts and Great Lakes provide specialized weather information for boaters, fishermen, and others engaged in marine activities, as well as general weather information.

During severe weather, National Weather Service forecasters can interrupt the routine weather broadcasts and substitute special warning messages. The forecasters can also activate specially designed warning receivers. Such receivers either sound an alarm indicating that an emergency exists, alerting the listener to turn the receiver up, or, when operated in a muted mode, automatically turns it up so the message can be heard. "Warning Alarm" receivers are especially valuable for schools, hospitals, public safety agencies, and news media offices.

NOAA Weather Radio broadcasts are made on one of seven high-band FM frequencies ranging from 162.40 to 162.55 megahertz (MHz). These frequencies are not found on the average home radio now in use. However, a number of radio man-ufacturers offer special weather radios to operate on these frequencies, with or without the emergency warning alarm. Also, there are now many radios on the market that offer standard AM/FM frequencies plus the so-called "weather band" as an added feature.

Table 7-4 offers a complete listing of the current stations in the NOAA Weather Radio Network. Figure 7-5 is a map illustrating locations of these stations. These broadcasts can usually be heard as far as 40 miles from the antenna site, sometimes more. The effective range depends on many factors, particularly the height of the broadcasting antenna, terrain, quality of the receiver, and type of receiving antenna. As a general rule, listeners close to or perhaps beyond the 40 mile range should have a good quality receiver system if they expect reliable reception. An outside antenna may also be needed in fringe areas. If practicable, the receiver should be tried at its place of intended use before making a final purchase.

About 90 percent of the nation's population is within listening range of a weather band broadcast.

Teletype

Aviators use the teletype as their source of weather information. It can also be utilized by other citizens to collect weather data (Fig. 7-6). The Federal Aviation Administration and the National Weather Service work together to supply and man aviation weather offices in airports across the nation. These offices are called Flight Service Stations, abbreviated FSS. These Flight Service Stations provide weather information to the public, but primarily to fliers, in person and over two-way radios.

Much of the FSS' weather information comes to them over teletypes from cities and stations across the nation. Figure 7-7 shows a typical "Terminal Forecast." A Terminal Forecast contains information for specific airports on ceiling, cloud heights, cloud amounts, visibility, weather and obstructions to vision, surface wind, and flight conditions. The terminal forecasts have a 24-hour valid period and are issued three times a day.

Area Forecasts are 18-hour forecasts plus a

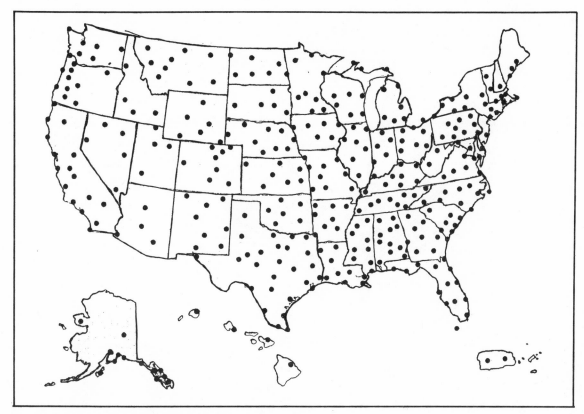

Fig. 7-5. NOAA National Weather Radio Network station locations.

12-hour categorical outlook prepared twice a day with general details of cloud, weather, and frontal conditions for an area the size of several states.

READING THE WEATHER MAP

The most common and useful indicator of present and future weather conditions is the weather map, also called the Surface Syntopic Weather Map. These maps, in various forms of complexity, are available through the government on a subscription basis (as mentioned earlier), in a simplified form in newspapers, or can be viewed at Flight Service Stations and National Weather Service offices. Figure 7-4 shows one of the these weather maps.

Many surface weather maps are prepared and transmitted each day by the National Meteorological Center (NMC). These maps are based on surface observed data recorded for a particular area.

They provide a map-oriented picture of the weather as it exists at the time of observation. This picture includes the atmospheric pressure patterns at the surface (pressure systems), fronts, and individual reporting station data.

Facsimile syntopic maps are drawn at the NMC eight times each day. The observation time appears in the lower left corner of the map. The map is drawn to a 1:10,000,000 scale and shows a great number of stations over the North American area. They are then transmitted either by facsimile transmission equipment or by computer to stations around the United States. A sequence of three or four maps is displayed in the weather office to indicate trends in the development of fronts, pressure systems, and weather phenomena (Table 7-5).

THE STATION MODEL

Data observed by individual stations are plot-

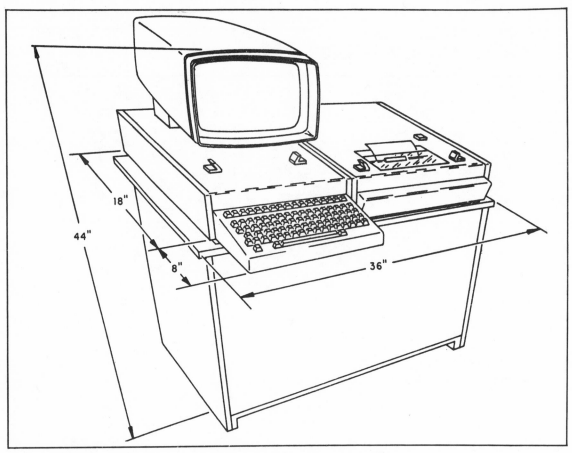

Fig. 7-6. VDT (Visual Display Terminal) used in many modern weather station offices.

```
ALABAMA 191442
HSV  191515 20SCT 250-BKN SCT V BKN. 17Z C30BKN CHC C100VC 2TRW.
   04Z 100SCT 5H. 09Z MVFR GF..
MOB  191515 10SCT 250-BKN 5H SCT V BKN. 17Z C30BKN CHC C80VC 1TRW.
   04Z 100SCT 250-BKN SLGT CHC RW/TRW. 09Z VFR BCMG MVFR GF..

GEORGIA 191442
ABY  191515 C8BKN. 17Z C35BKN 0910 BKN V SCT CHC C8X 1/2TRW. 02Z CLR.
   08Z C5BKN CHC C2X 1/4F. 09Z IFR CIG F..
AGS  191515 C15BKN. 17Z C35BKN 1210 BKN V SCT CHC C8X 1/2TRW. 02Z CLR.
   06Z C5BKN CHC C2X 1/4F. 09Z IFR CIG F..
AHN  191515 C15BKN. 17Z C35BKN 0710 BKN V SCT CHC C8X 1/2TRW. 02Z CLR.
   06Z C5BKN CHC C2X 1/4F. 09Z IFR CIG F..
MCN  191515 C8BKN. 17Z C35BKN 0910 BKN V SCT CHC C8X 1/2TRW. 02Z CLR.
   07Z C5BKN CHC C2X 1/4F. 09Z IFR CIG F..
SAV  191515 15SCT C30BKN 0910 LYRS SCT V BKN OCNL C8X 1/2RW/TRW.
   09Z MVFR CIG RW..
```

Fig. 7-7. Teletyped terminal forecast.

Table 7-5. Color Codes for Precipitation on Weather Maps.

Precipitation areas	Coloring
Areas of continuous precipitation.	Light green shading.
Areas of intermittent precipitation.	Light green hatching.
Areas of showers.	Green shower symbols.
Areas of drizzle.	Green drizzle symbols.
Areas of past thunderstorms.	Red thunderstorm symbols.
Areas of present thunderstorms.	Green thunderstorm symbols.
Areas of fog.	Light yellow shading.

Fig. 7-8. Syntopic station model.

ted in the form of standard symbols on the surface weather map. These symbols are grouped around a printed circle called the *station circle,* which indicates the geographical location of the station on the chart (Fig. 7-8). This grouping is the *station model.* The arrangement of plotted data around the station model is standard in weather stations throughout the world. On NMC facsimile maps, the abbreviated station model includes only the sky cover, surface wind, present weather, temperature, dew point, cloud types, pressure, pressure tendency, and accumulated precipitation.

The direction from which the wind is blowing is indicated on the surface weather map by a shaft.

The shaft on the station models in Fig. 7-9 indicates wind from the northwest. Wind speed is indicated by the *barb and pennant* system—a full barb is valued at 10 knots, a half-barb at 5 knots, and a pennant at 50 knots. To avoid confusion in length with a single full barb, the half-barb appears part way down the shaft when 5-knot winds exist. The wind speed is a summation of the number of barbs and pennants appearing on the shaft. For a calm wind, a circle is drawn around the station center. The wind direction is determined by reference to true north (a true wind), and the speed is rounded to the nearest 5 knots.

The amount of black in the fascimile station

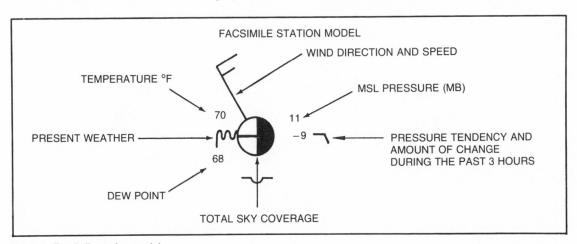

Fig. 7-9. Facsimile station model.

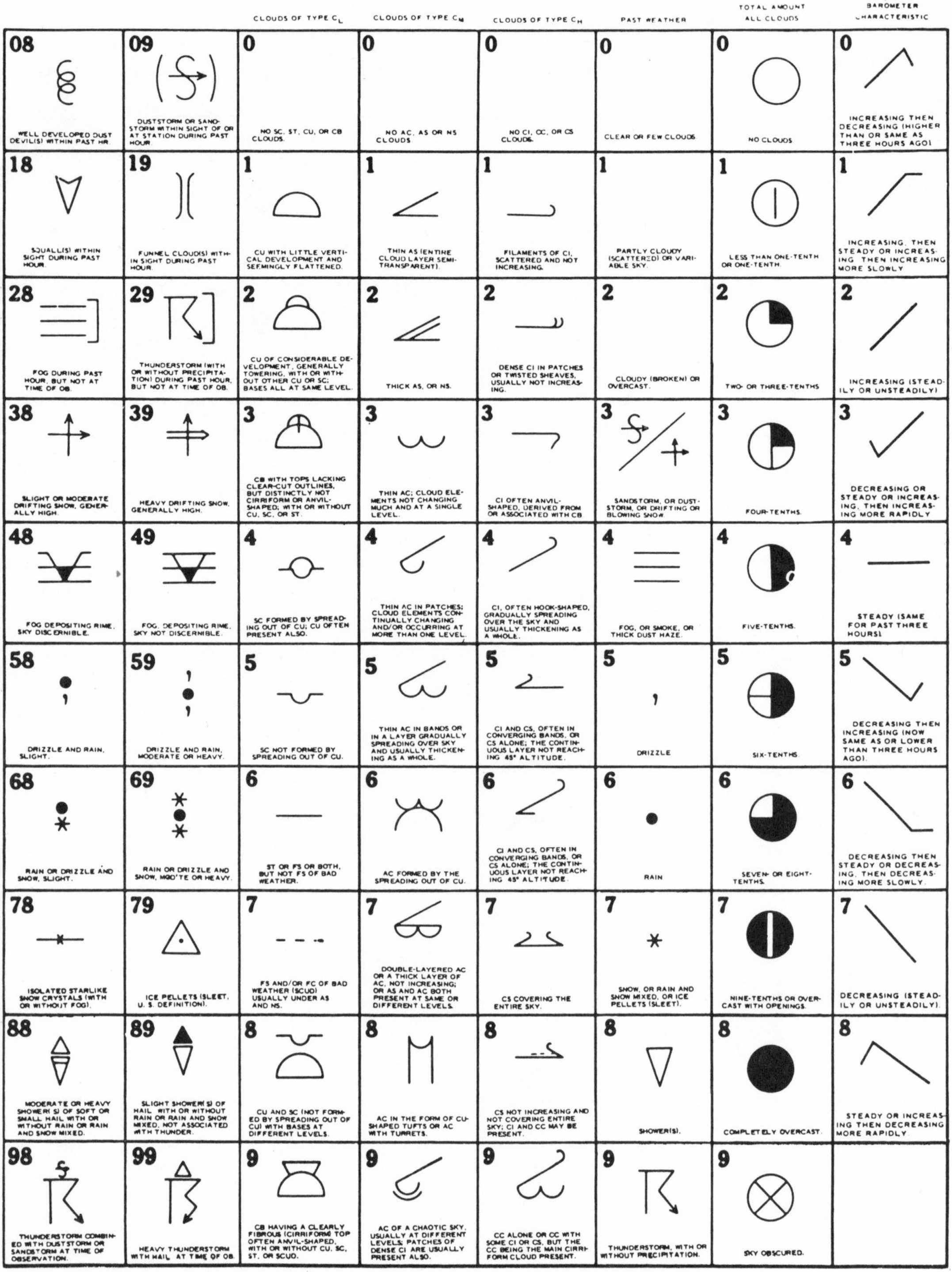
CLOUDS OF TYPE C$_L$
CLOUDS OF TYPE C$_M$
CLOUDS OF TYPE C$_H$
PAST WEATHER
TOTAL AMOUNT ALL CLOUDS
BAROMETER CHARACTERISTIC
08 WELL DEVELOPED DUST DEVIL(S) WITHIN PAST HR.
09 DUSTSTORM OR SANDSTORM WITHIN SIGHT OF OR AT STATION DURING PAST HOUR.
0 NO SC, ST, CU, OR CB CLOUDS.
0 NO AC, AS OR NS CLOUDS
0 NO CI, CC, OR CS CLOUDS.
0 CLEAR OR FEW CLOUDS.
0 NO CLOUDS
0 INCREASING THEN DECREASING (HIGHER THAN OR SAME AS THREE HOURS AGO)
18 SQUALL(S) WITHIN SIGHT DURING PAST HOUR.
19 FUNNEL CLOUD(S) WITHIN SIGHT DURING PAST HOUR.
1 CU WITH LITTLE VERTICAL DEVELOPMENT AND SEEMINGLY FLATTENED.
1 THIN AS (ENTIRE CLOUD LAYER SEMI-TRANSPARENT).
1 FILAMENTS OF CI, SCATTERED AND NOT INCREASING.
1 PARTLY CLOUDY (SCATTERED) OR VARIABLE SKY.
1 LESS THAN ONE-TENTH OR ONE-TENTH.
1 INCREASING, THEN STEADY OR INCREASING THEN INCREASING MORE SLOWLY
28 FOG DURING PAST HOUR, BUT NOT AT TIME OF OB.
29 THUNDERSTORM (WITH OR WITHOUT PRECIPITATION) DURING PAST HOUR, BUT NOT AT TIME OF OB.
2 CU OF CONSIDERABLE DEVELOPMENT, GENERALLY TOWERING, WITH OR WITHOUT OTHER CU OR SC; BASES ALL AT SAME LEVEL.
2 THICK AS, OR NS.
2 DENSE CI IN PATCHES OR TWISTED SHEAVES, USUALLY NOT INCREASING.
2 CLOUDY (BROKEN) OR OVERCAST.
2 TWO- OR THREE-TENTHS
2 INCREASING (STEADILY OR UNSTEADILY)
38 SLIGHT OR MODERATE DRIFTING SNOW, GENERALLY HIGH.
39 HEAVY DRIFTING SNOW, GENERALLY HIGH.
3 CB WITH TOPS LACKING CLEAR-CUT OUTLINES, BUT DISTINCTLY NOT CIRRIFORM OR ANVIL-SHAPED; WITH OR WITHOUT CU, SC, OR ST.
3 THIN AC; CLOUD ELEMENTS NOT CHANGING MUCH AND AT A SINGLE LEVEL.
3 CI OFTEN ANVIL-SHAPED, DERIVED FROM OR ASSOCIATED WITH CB
3 SANDSTORM, OR DUSTSTORM, OR DRIFTING OR BLOWING SNOW.
3 FOUR-TENTHS.
3 DECREASING OR STEADY OR INCREASING, THEN INCREASING MORE RAPIDLY
48 FOG DEPOSITING RIME, SKY DISCERNIBLE.
49 FOG, DEPOSITING RIME, SKY NOT DISCERNIBLE.
4 SC FORMED BY SPREADING OUT OF CU; CU OFTEN PRESENT ALSO.
4 THIN AC IN PATCHES; CLOUD ELEMENTS CONTINUALLY CHANGING AND/OR OCCURRING AT MORE THAN ONE LEVEL.
4 CI, OFTEN HOOK-SHAPED, GRADUALLY SPREADING OVER THE SKY AND USUALLY THICKENING AS A WHOLE.
4 FOG, OR SMOKE, OR THICK DUST HAZE.
4 FIVE-TENTHS.
4 STEADY (SAME FOR PAST THREE HOURS)
58 DRIZZLE AND RAIN, SLIGHT.
59 DRIZZLE AND RAIN, MODERATE OR HEAVY.
5 SC NOT FORMED BY SPREADING OUT OF CU.
5 THIN AC IN BANDS OR IN A LAYER GRADUALLY SPREADING OVER SKY AND USUALLY THICKENING AS A WHOLE.
5 CI AND CS, OFTEN IN CONVERGING BANDS, OR CS ALONE; THE CONTINUOUS LAYER NOT REACHING 45° ALTITUDE.
5 DRIZZLE
5 SIX-TENTHS.
5 DECREASING THEN INCREASING (NOW SAME AS OR LOWER THAN THREE HOURS AGO).
68 RAIN OR DRIZZLE AND SNOW, SLIGHT.
69 RAIN OR DRIZZLE AND SNOW, MOD'TE OR HEAVY.
6 ST OR FS OR BOTH, BUT NOT FS OF BAD WEATHER.
6 AC FORMED BY THE SPREADING OUT OF CU.
6 CI AND CS, OFTEN IN CONVERGING BANDS, OR CS ALONE; THE CONTINUOUS LAYER NOT REACHING 45° ALTITUDE.
6 RAIN
6 SEVEN- OR EIGHT-TENTHS.
6 DECREASING THEN STEADY OR DECREASING, THEN DECREASING MORE SLOWLY.
78 ISOLATED STARLIKE SNOW CRYSTALS (WITH OR WITHOUT FOG).
79 ICE PELLETS (SLEET, U. S. DEFINITION).
7 FS AND/OR FC OF BAD WEATHER (SCUD) USUALLY UNDER AS AND NS.
7 DOUBLE-LAYERED AC OR A THICK LAYER OF AC, NOT INCREASING; OR AS AND AC BOTH PRESENT AT SAME OR DIFFERENT LEVELS.
7 CS COVERING THE ENTIRE SKY.
7 SNOW, OR RAIN AND SNOW MIXED, OR ICE PELLETS (SLEET).
7 NINE-TENTHS OR OVERCAST WITH OPENINGS.
7 DECREASING (STEADILY OR UNSTEADILY).
88 MODERATE OR HEAVY SHOWER(S) OF SOFT OR SMALL HAIL WITH OR WITHOUT RAIN OR RAIN AND SNOW MIXED.
89 SLIGHT SHOWER(S) OF HAIL WITH OR WITHOUT RAIN OR RAIN AND SNOW MIXED, NOT ASSOCIATED WITH THUNDER.
8 CU AND SC (NOT FORMED BY SPREADING OUT OF CU) WITH BASES AT DIFFERENT LEVELS.
8 AC IN THE FORM OF CU-SHAPED TUFTS OR AC WITH TURRETS.
8 CS NOT INCREASING AND NOT COVERING ENTIRE SKY; CI AND CC MAY BE PRESENT.
8 SHOWER(S).
8 COMPLETELY OVERCAST.
8 STEADY OR INCREASING THEN DECREASING MORE RAPIDLY
98 THUNDERSTORM COMBINED WITH DUSTSTORM OR SANDSTORM AT TIME OF OBSERVATION.
99 HEAVY THUNDERSTORM WITH HAIL AT TIME OF OB.
9 CB HAVING A CLEARLY FIBROUS (CIRRIFORM) TOP OFTEN ANVIL-SHAPED, WITH OR WITHOUT CU, SC, ST, OR SCUD.
9 AC OF A CHAOTIC SKY, USUALLY AT DIFFERENT LEVELS; PATCHES OF DENSE CI ARE USUALLY PRESENT ALSO.
9 CC ALONE OR CC WITH SOME CI OR CS, BUT THE CC BEING THE MAIN CIRRIFORM CLOUD PRESENT.
9 THUNDERSTORM, WITH OR WITHOUT PRECIPITATION.
9 SKY OBSCURED.

Fig. 7-10. Reading the daily weather map.

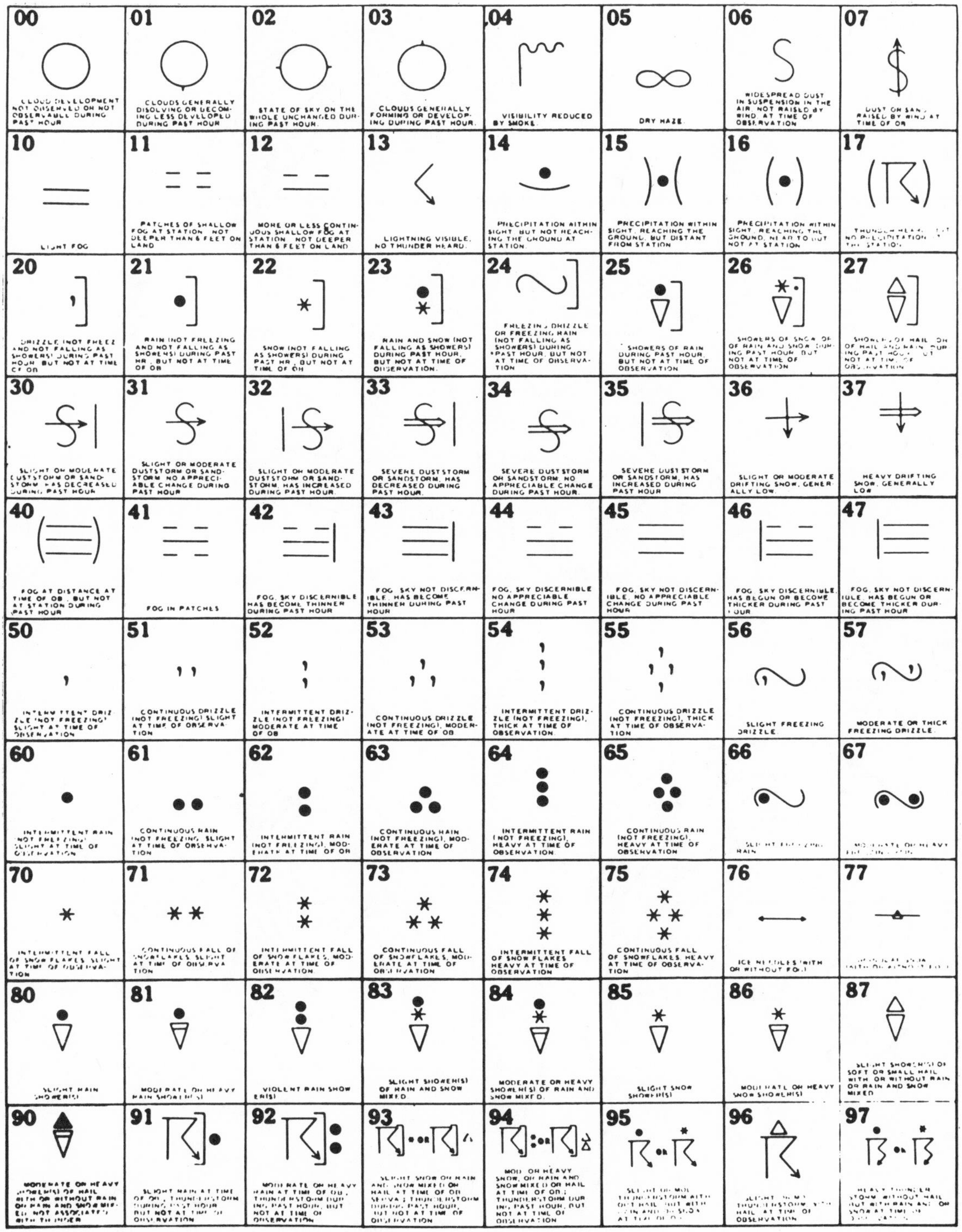

Code	Description
00	CLOUD DEVELOPMENT NOT OBSERVED OR NOT OBSERVABLE DURING PAST HOUR
01	CLOUDS GENERALLY DISSOLVING OR BECOMING LESS DEVELOPED DURING PAST HOUR
02	STATE OF SKY ON THE WHOLE UNCHANGED DURING PAST HOUR
03	CLOUDS GENERALLY FORMING OR DEVELOPING DURING PAST HOUR
04	VISIBILITY REDUCED BY SMOKE
05	DRY HAZE
06	WIDESPREAD DUST IN SUSPENSION IN THE AIR (NOT RAISED BY WIND) AT TIME OF OBSERVATION
07	DUST OR SAND RAISED BY WIND AT TIME OF OB
10	LIGHT FOG
11	PATCHES OF SHALLOW FOG AT STATION NOT DEEPER THAN 6 FEET ON LAND
12	MORE OR LESS CONTINUOUS SHALLOW FOG AT STATION NOT DEEPER THAN 6 FEET ON LAND
13	LIGHTNING VISIBLE, NO THUNDER HEARD
14	PRECIPITATION WITHIN SIGHT BUT NOT REACHING THE GROUND AT STATION
15	PRECIPITATION WITHIN SIGHT, REACHING THE GROUND, BUT DISTANT FROM STATION
16	PRECIPITATION WITHIN SIGHT, REACHING THE GROUND, NEAR TO BUT NOT AT STATION
17	THUNDER HEARD, BUT NO PRECIPITATION AT THE STATION
20	DRIZZLE (NOT FREEZING AND NOT FALLING AS SHOWERS) DURING PAST HOUR, BUT NOT AT TIME OF OB
21	RAIN (NOT FREEZING AND NOT FALLING AS SHOWERS) DURING PAST HR., BUT NOT AT TIME OF OB
22	SNOW (NOT FALLING AS SHOWERS) DURING PAST HR., BUT NOT AT TIME OF OB
23	RAIN AND SNOW (NOT FALLING AS SHOWERS) DURING PAST HR., BUT NOT AT TIME OF OBSERVATION
24	FREEZING DRIZZLE OR FREEZING RAIN (NOT FALLING AS SHOWERS) DURING PAST HOUR, BUT NOT AT TIME OF OBSERVATION
25	SHOWERS OF RAIN DURING PAST HOUR BUT NOT AT TIME OF OBSERVATION
26	SHOWERS OF SNOW OR OF RAIN AND SNOW DURING PAST HOUR BUT NOT AT TIME OF OBSERVATION
27	SHOWERS OF HAIL OR OF HAIL AND RAIN DURING PAST HR., BUT NOT AT TIME OF OBSERVATION
30	SLIGHT OR MODERATE DUSTSTORM OR SANDSTORM HAS DECREASED DURING PAST HOUR
31	SLIGHT OR MODERATE DUSTSTORM OR SANDSTORM NO APPRECIABLE CHANGE DURING PAST HOUR
32	SLIGHT OR MODERATE DUSTSTORM OR SANDSTORM HAS INCREASED DURING PAST HOUR
33	SEVERE DUSTSTORM OR SANDSTORM, HAS DECREASED DURING PAST HOUR
34	SEVERE DUSTSTORM OR SANDSTORM, NO APPRECIABLE CHANGE DURING PAST HOUR
35	SEVERE DUSTSTORM OR SANDSTORM, HAS INCREASED DURING PAST HOUR
36	SLIGHT OR MODERATE DRIFTING SNOW, GENERALLY LOW
37	HEAVY DRIFTING SNOW, GENERALLY LOW
40	FOG AT DISTANCE AT TIME OF OB., BUT NOT AT STATION DURING PAST HOUR
41	FOG IN PATCHES
42	FOG, SKY DISCERNIBLE HAS BECOME THINNER DURING PAST HOUR
43	FOG, SKY NOT DISCERNIBLE HAS BECOME THINNER DURING PAST HOUR
44	FOG, SKY DISCERNIBLE NO APPRECIABLE CHANGE DURING PAST HOUR
45	FOG, SKY NOT DISCERNIBLE NO APPRECIABLE CHANGE DURING PAST HOUR
46	FOG, SKY DISCERNIBLE HAS BEGUN OR BECOME THICKER DURING PAST HOUR
47	FOG, SKY NOT DISCERNIBLE HAS BEGUN OR BECOME THICKER DURING PAST HOUR
50	INTERMITTENT DRIZZLE (NOT FREEZING) SLIGHT AT TIME OF OBSERVATION
51	CONTINUOUS DRIZZLE (NOT FREEZING) SLIGHT AT TIME OF OBSERVATION
52	INTERMITTENT DRIZZLE (NOT FREEZING) MODERATE AT TIME OF OB
53	CONTINUOUS DRIZZLE (NOT FREEZING) MODERATE AT TIME OF OB
54	INTERMITTENT DRIZZLE (NOT FREEZING) THICK AT TIME OF OBSERVATION
55	CONTINUOUS DRIZZLE (NOT FREEZING) THICK AT TIME OF OBSERVATION
56	SLIGHT FREEZING DRIZZLE
57	MODERATE OR THICK FREEZING DRIZZLE
60	INTERMITTENT RAIN (NOT FREEZING) SLIGHT AT TIME OF OBSERVATION
61	CONTINUOUS RAIN (NOT FREEZING) SLIGHT AT TIME OF OBSERVATION
62	INTERMITTENT RAIN (NOT FREEZING) MODERATE AT TIME OF OB
63	CONTINUOUS RAIN (NOT FREEZING) MODERATE AT TIME OF OBSERVATION
64	INTERMITTENT RAIN (NOT FREEZING) HEAVY AT TIME OF OBSERVATION
65	CONTINUOUS RAIN (NOT FREEZING) HEAVY AT TIME OF OBSERVATION
66	SLIGHT FREEZING RAIN
67	MODERATE OR HEAVY FREEZING RAIN
70	INTERMITTENT FALL OF SNOWFLAKES SLIGHT AT TIME OF OBSERVATION
71	CONTINUOUS FALL OF SNOWFLAKES SLIGHT AT TIME OF OBSERVATION
72	INTERMITTENT FALL OF SNOWFLAKES MODERATE AT TIME OF OBSERVATION
73	CONTINUOUS FALL OF SNOWFLAKES MODERATE AT TIME OF OBSERVATION
74	INTERMITTENT FALL OF SNOWFLAKES HEAVY AT TIME OF OBSERVATION
75	CONTINUOUS FALL OF SNOWFLAKES HEAVY AT TIME OF OBSERVATION
76	ICE NEEDLES (WITH OR WITHOUT FOG)
77	GRANULAR SNOW (WITH OR WITHOUT FOG)
80	SLIGHT RAIN SHOWER(S)
81	MODERATE OR HEAVY RAIN SHOWER(S)
82	VIOLENT RAIN SHOWER(S)
83	SLIGHT SHOWER(S) OF RAIN AND SNOW MIXED
84	MODERATE OR HEAVY SHOWER(S) OF RAIN AND SNOW MIXED
85	SLIGHT SNOW SHOWER(S)
86	MODERATE OR HEAVY SNOW SHOWER(S)
87	SLIGHT SHOWER(S) OF SOFT OR SMALL HAIL WITH OR WITHOUT RAIN OR RAIN AND SNOW MIXED
90	MODERATE OR HEAVY SHOWER(S) OF HAIL WITH OR WITHOUT RAIN OR RAIN AND SNOW MIXED ASSOCIATED WITH THUNDER
91	SLIGHT RAIN AT TIME OF OB., THUNDERSTORM DURING PAST HOUR BUT NOT AT TIME OF OBSERVATION
92	MODERATE OR HEAVY RAIN AT TIME OF OB., THUNDERSTORM DURING PAST HOUR BUT NOT AT TIME OF OBSERVATION
93	SLIGHT SNOW OR RAIN AND SNOW MIXED OR HAIL AT TIME OF OB., THUNDERSTORM DURING PAST HOUR, BUT NOT AT TIME OF OBSERVATION
94	MODERATE OR HEAVY SNOW, OR RAIN AND SNOW MIXED OR HAIL AT TIME OF OB., THUNDERSTORM DURING PAST HOUR, BUT NOT AT TIME OF OBSERVATION
95	SLIGHT OR MODERATE THUNDERSTORM WITHOUT HAIL BUT WITH RAIN AND/OR SNOW AT TIME OF OBSERVATION
96	SLIGHT OR MODERATE THUNDERSTORM WITH HAIL AT TIME OF OBSERVATION
97	HEAVY THUNDERSTORM WITHOUT HAIL BUT WITH RAIN AND/OR SNOW AT TIME OF OBSERVATION

Fig. 7-11. Reading the daily weather map (continued).

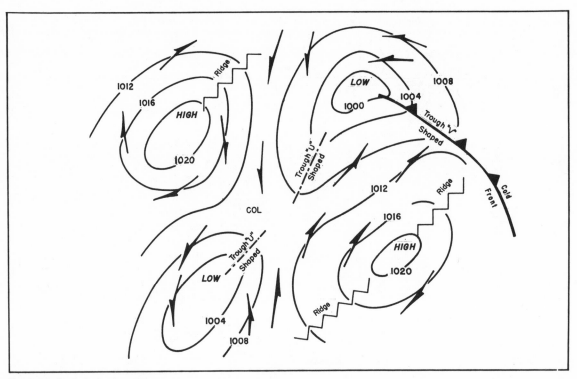

Fig. 7-12. Basic isobaric patterns.

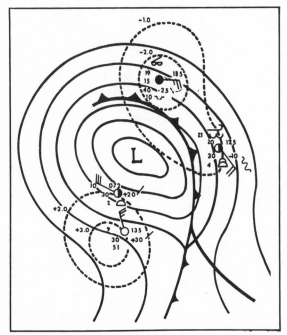

Fig. 7-13. Isallobars.

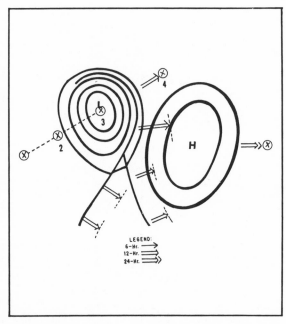

Fig. 7-14. Forecasting the movement of fronts and pressure systems.

circle is an indication of the total amount of sky coverage by all cloud layers over the station (Fig. 7-9). These symbols are part of the international surface syntopic code.

The mean sea level pressure is indicated to the nearest millibar (two digits). If the reported pressure begins with a digit from 0 through 5, the figure is preceded with a 10, (i.e.), 99 means 999 millibars and 65 means 965 millibars. These pressure values are corrected to mean sea level to eliminate pressure variations caused by varying field elevations (Figs. 7-12 through 7-14).

On surface weather maps prepared by the United States weather services the temperatures and dew points are given in degrees Fahrenheit. However, weather maps produced in most other major countries are given in degrees Celsius on the centigrade scale.

The symbols used for present weather, illustrated in Figs. 7-10 and 7-11 represent forms of precipitation or obstructions to vision, or both. Although there are 100 possible combinations and types of weather symbols, only a few are commonly used (Fig. 7-2).

Pressure tendency changes over the past three hours are also indicated. A plus or minus indicates whether the net change in pressure has resulted in an increase or decrease. The amount of change is indicated in tenths of millibars, and the pattern of change is indicated by the coded symbol.

Although there are many cloud types symbolized in the international code, a knowledge of the

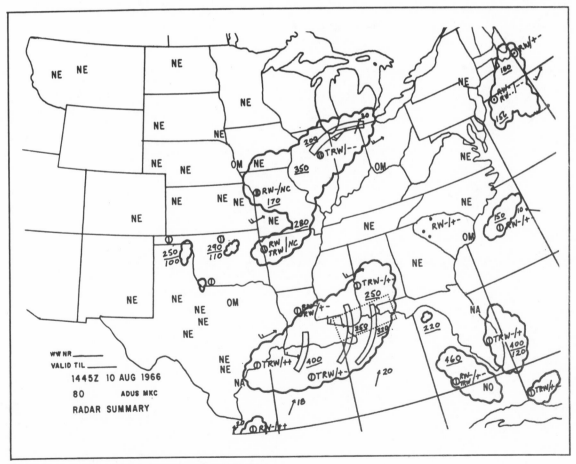

Fig. 7-15. Example of a Radar Summary facsimile chart.

ten basic types will be adequate. The same symbol, plotted above the station circle, is used for nimbostratus or thick altostratus clouds; the two cloud types are similar in that both may obscure the sun and produce continuous precipitation at the surface.

Once collected and recorded, weather data is interpreted following the rules of meteorology (Fig. 7-15). These forecasts of future weather conditions can then be used to benefit pilots, farmers, business, students, parents, and citizens.

Chapter 8

Interpreting Weather Data

Once weather information is gathered it must be interpreted in order to be useful. Weather data may tell you that a high pressure system is moving in your direction at approximately 75 miles a day. It is now about 200 miles from you and is nearly 300 miles wide. This data can be *interpreted* to say that in about two and a half days you can expect fair weather and warmer temperatures lasting approximately four days. To *use* the weather information, you might plan a beach outing during the time the high is in your area.

The interpretation of weather data is as much a science as collecting weather information—though not as exact. Still, the rules of understanding and translating weather data into useful forecasts can easily be applied by the amateur meteorologist.

Weather is caused by the movement of the atmosphere as the Earth rotates around the sun. More specifically, it is caused by the movement of systems or areas of high and low pressure. Chapter 2 gave you a background of how air masses, pressure systems, and fronts are formed and move.

By keeping the weather maps that appear in the daily newspaper and by watching the local tele-vision meteorologists, you can keep accurate track of the surface pressure systems that move into, through, and out of your local area (Fig. 8-1). To better understand and forecast local weather, let's consider the air masses that affect weather in the continental United States.

MARITIME POLAR AIR MASSES

Maritime polar air masses that invade the U.S. arrive from two different source regions. One of these is located in the North Pacific Ocean, the other in the northwestern portion of the North Atlantic Ocean. Those air masses orginating over the Pacific Ocean dominate the weather conditions of the Pacific coast of the United States and western Canada. Those air masses originating over the North Atlantic Ocean frequently appear during the winter over the northeastern coast of the U.S.

Many of the winter maritime polar air masses that invade the Pacific coast originate in the interior of Siberia. They have a long overwater trajectory and, during this travel over the Pacific Ocean, are unstable in the lower layers (Fig. 8-2). As they

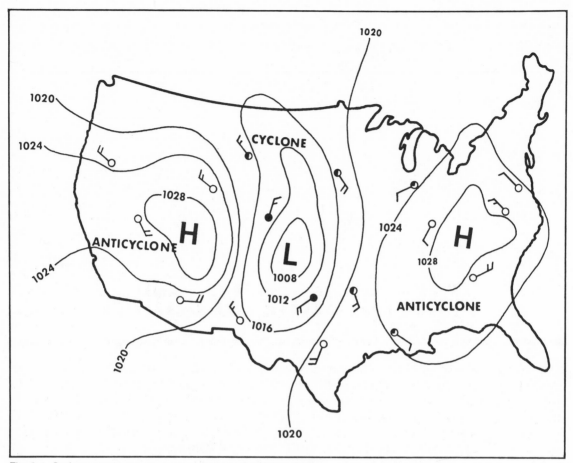

Fig. 8-1. Surface windflow in pressure systems in the United States.

invade the west coast, they are cooled from below by a cool ocean current and coastal area, and become more stable. Along the Pacific coastal regions, stratus and stratocumulus clouds are common in these air masses. Maritime polar air masses cause heavy cumuliform cloud formation and extensive shower activity as they move eastward up the slopes of the mountains. East of the mountains, the air descends and is warmed adiabatically (by result of increasing pressure). This adiabatic warming results in decreased relative humidity and the skies are generally clear.

In the northeastern section of the United States, maritime polar air moves into the New England States from the northeast. These air masses are usually colder and more stable than those en-

tering the west coast from a northwest direction. Low stratiform clouds with light continuous precipitation and generally strong winds occur as these air masses move inland.

Since water temperatures are cooler than adjacent land temperatures in the summer, maritime polar air masses entering the Pacific coast become unstable because of the surface heating. In the afternoon, cumuliform cloud formations and widely scattered showers occur. At night, fog and low stratiform clouds are common on the coastal regions, especially along the coast of California. When the air masses cross the mountains, they lose a considerable amount of moisture on the western mountain slopes. The orographic lifting intensifies the development of cumuliform clouds on the windward slopes. These cloud buildups are accompanied by heavy showers.

CONTINENTAL POLAR AIR MASSES

Continental polar air masses that invade the United States during winter originate over Canada and Alaska. They are stable in the source regions. As the air masses move southward into the U.S., they are heated by the underlying surface. During daylight hours, the air is generally unstable near the surface and the sky is usually clear. At night the air tends to become more stable. When these cold, dry air masses move over the warmer waters of the Great Lakes, they acquire heat and moisture and become unstable in the lower levels. Cumuliform clouds form and produce snow flurries over the Great Lakes and on the leeward side of the lakes. As the air masses move southeastward, the cumu-

liform clouds intensify along the Appalachian Mountains. Continental polar air masses between the Great Lakes and the peaks of the Appalachians contain some of the most unfavorable flying conditions in the United States during the winter months. Clear skies or scattered clouds are normal east of the mountains.

Cold dry air masses have different characteristics and properties in the summer than in the winter. Since the thawed-out source regions are warmer and contain more moisture, the air is less stable in the surface layers. The air is, therefore, cool and contains slightly more moisture when it reaches the United States. Scattered cumuliform clouds form during the day in this unstable air, but dissipate at night when the air becomes more stable. When these air masses move over the colder water of the Great Lakes in the summer, they are

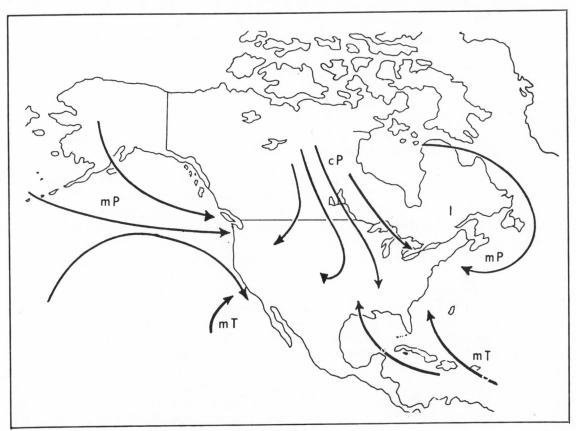

Fig. 8-2. Winter air mass source regions and paths in North America.

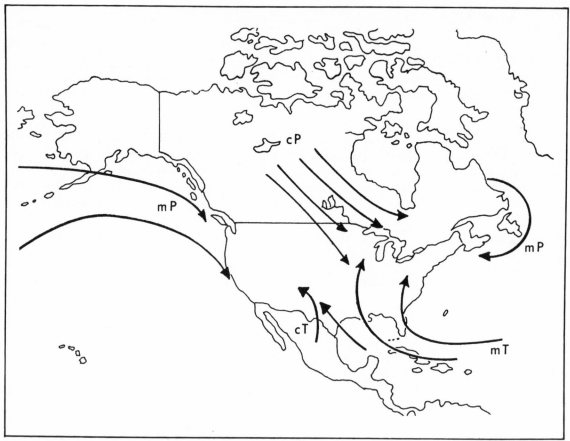

Fig. 8-3. Summer air mass source regions and paths in North America.

cooled from below and become stable, resulting in good weather.

MARITIME TROPICAL AIR MASSES

Maritime tropical air masses originate over the Atlantic Ocean, the Gulf of Mexico, and the Caribbean Sea. They move into the U.S. from the Gulf of Mexico or Atlantic Ocean and are common along the Southeastern and Gulf Coast States (Fig. 8-3). Warm, moist, stable air that originates over the Pacific Ocean is rarely observed in Southwestern United States because the prevailing winds in the southwest blow offshore.

Because the land is colder in winter than the water, warm, moist air masses are cooled from below and become stable as they move inland over the South Atlantic and Gulf States. Fog and strati-form clouds form at night over the coastal regions. The fog and clouds tend to dissipate or become stratocumulus during the afternoon. The extent to which the cloudiness and fog spread inland is dependent on the difference between the surface temperature and the air temperature. When surface temperatures are cold, fog and stratiform clouds extend inland for considerable distances throughout the Eastern States. When land temperatures are extremely cold, extensive surface temperature inversions develop. Under such conditions, daytime heating usually does not eliminate the inversions and the fog and stratiform clouds may persist for several days. In winter, when the air moves over the Appalachian Mountains, the adiabatic cooling produced by orographic lifting causes heavy cumuliform clouds to form on the windward side.

Warm moist air covers the eastern half of the United States during most of the summer. Since the land is normally warmer than the water, particularly during the day, this air mass is heated from below by the surface and becomes unstable as it moved inland. Along the coastal regions, stratiform clouds are common during the early morning hours. These stratiform clouds usually change during the later morning to scattered cumuliform clouds. By late afternoon, extensive areas of widely scattered thunderstorms normally develop. In maritime tropical air masses, cumuliform clouds and thunderstorms are usually more numerous and intense on the windward side of mountain ranges, in squall lines and in prefrontal activity.

CONTINENTAL TROPICAL AIR MASSES

Continental tropical air masses are observed primarily in the Mexico/Texas/Arizona/New Mexico area (their "source region") and only in the summer. These air masses are characterized by high temperatures, low humidity and, although extremely rare, scattered cumuliform clouds. The bases of these clouds are exceptionally high, causing turbulent air below them.

PLOTTING THE WEATHER

Once you are aware of the typical movement of air masses in your region of the country, you can better understand and predict the movement of fronts and associated weather (Fig. 8-4). In the coming pages you'll learn how compiled data and weather norms can make your forecasts as accurate as the weatherman's.

First, remember that most weather in the United States moves from west to east due to the rotation of the earth (Fig. 8-5). These weather systems move at average speeds of 15 miles an hour in the summer to 25 miles an hour in winter months, carried along by air streams called the *prevailing westerlies*. Information gathered on barometric pressure will tell you whether you are now within a high pressure (above 1,016 millibars or 30 inches of mercury) system or a low pressure (below 1,016 millibars) system. Barometric pressure that is ris-

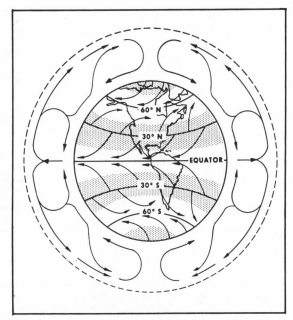

Fig. 8-4. Three-cell theory of circulation.

ing or falling can tell you whether the center of the pressure system is coming closer or moving away from your location. Logic tells us that pressure above 1,016 and rising means the high pressure system is coming towards us. Falling pressure means it is moving away. Pressure readings below 1,016 and falling mean a low pressure system is coming in. Rising pressure would mean it has passed.

You can also check the wind to confirm what type of pressure system is predominant. As you remember from our discussion of pressure systems, wind flows more or less with the isobars (lines connecting areas of same barometric pressure) *clockwise* around a *high* and *counterclockwise* around a *low*. A simple way of finding high and low systems in your area is to turn your back on the wind in your area, then point to the left toward the low pressure system and right for the high pressure system (Figs. 8-6 through 8-9).

Another generalization is that highs often bring good weather and lows bring bad weather.

You can then plot the weather by using newspaper or television weather maps to compare where a system was yesterday and where it is today. Note

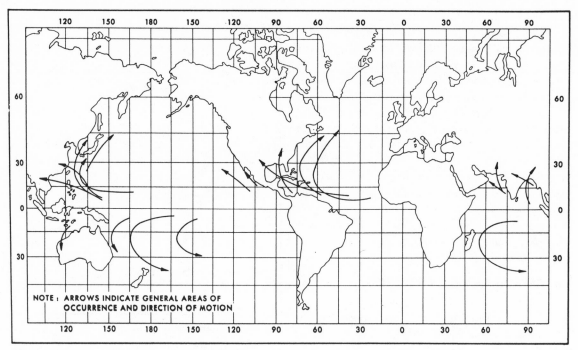

Fig. 8-5. Principal world regions of tropical cyclones.

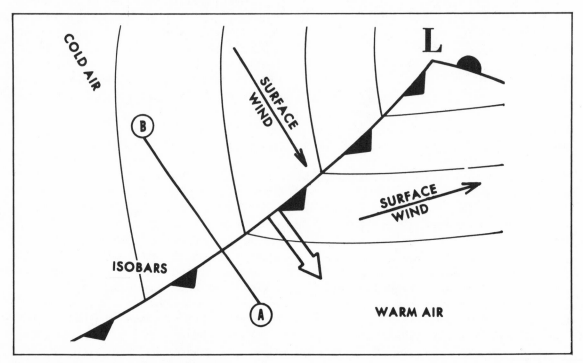

Fig. 8-6. Cold front on a surface weather map.

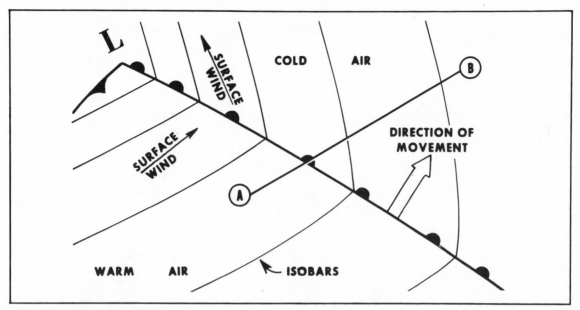

Fig. 8-7. Warm front on a surface weather map.

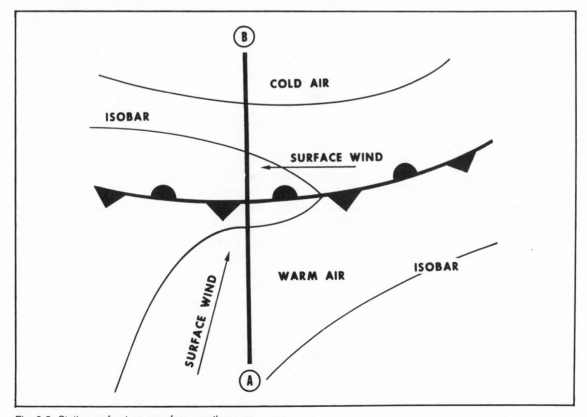

Fig. 8-8. Stationary front on a surface weather map.

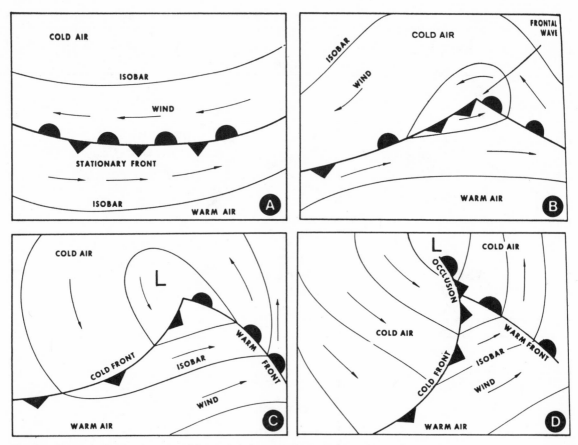

Fig. 8-9. Stages in the development of the occluded wave on a surface weather map.

how many miles the system has moved in the past 24 hours, then project where the system will be tomorrow. Keep two things in mind as you plot: first, that the earth rotates, spinning systems to the left or right of a straight path depending on the type of system; second, most maps are flat maps rather than round and tend to distort the actual distances and paths that systems seem to take.

FORECASTING YOUR OWN WEATHER

Since these high and low pressure systems bring in the weather to our area, the ability to read incoming systems in advance will certainly help us predict tomorrow's weather. And, as we have just seen, two excellent indicators of future systems are the barometric pressure (highs and lows) and the wind direction (location and path of the system). If you compiled a chart of wind direction and barometric pressure relationships, that information could be used to forecast weather on a short-term basis.

Table 8-1 is such a chart, useful for many parts of the United States. With this chart, a weathervane, and a barometer you can predict weather for up to 48 hours in advance with a high degree of accuracy.

For example, your barometer reads 30.10 to 30.20 inches of mercury and falling slowly while the wind comes from the south or southeast. You can expect rain within 24 hours. Why? Because a low pressure system is moving in, pushing out the high.

You can also confirm your barometric/wind forecasts by reading the clouds. Clouds mark the circulation of the lower atmosphere, trailing out along the bands of wind at various levels, or growing with vertical movement of the air. Because clouds are conglomerations of condensed liquid or frozen water, their visible characteristics also reflect the distribution of temperature and moisture content of the atmosphere in their immediate vicinity.

The cirrus or ice crystal cloud, for example, is generally followed by low and middle clouds; a cross-sectional view along this parade of clouds would show them forming along a slope, ranging from cirrus at the high end to altostratus or stratus at the other. The persistence of turbulence—in-

stability within air masses—is also indicated by the type of cloud. Approaching cirrocumulus clouds hearld the arrival of unstable air, and the development from fair-weather cumulus clouds to cumulonimbus giants is a familiar sight.

To generalize, improving weather conditions are indicated by steady decrease in the number of clouds, higher cloud bases, increasing breaks in the overcast, and fog dissipating before noon. The changes indicating poor weather increase if fast-moving clouds thicken and lower, a line of middle clouds darken on the western horizon, isolated roll

Table 8-1. Simple Wind/Pressure Weather Forecasting Chart.

Wind direction	Barometer reduced to sea level	Character of weather indicated
SW. to NW	30.10 to 30.20 and steady	Fair, with slight temperature changes for 1 to 2 days.
SW. to NW	30.10 to 30.20 and rising rapidly	Fair, followed within 2 days by rain.
SW. to NW	30.20 and above and stationary	Continued fair, with no decided temperature change.
SW. to NW	30.20 and above and falling slowly	Slowly rising temperature and fair for 2 days.
S. to SE	30.10 to 30.20 and falling slowly	Rain within 24 hours.
S. to SE	30.10 to 30.20 and falling rapidly	Wind increasing in force, with rain within 12 to 24 hours.
SE. to NE	30.10 to 30.20 and falling slowly	Rain in 12 to 18 hours.
SE. to NE	30.10 to 30.20 and falling rapidly	Increasing wind, and rain within 12 hours.
E. to NE	30.10 and above and falling slowly	In summer, with light winds, rain may not fall for several days. In winter, rain within 24 hours.
E. to NE	30.10 and above and falling rapidly	In summer, rain probable within 12 to 24 hours. In winter, rain or snow, with increasing winds, will often set in when the barometer begins to fall and the wind sets in from the NE.
SE. to NE	30.00 or below and falling slowly	Rain will continue 1 to 2 days.
SE. to NE	30.00 or below and falling rapidly	Rain, with high wind, followed, within 36 hours, by clearing, and in winter by colder.
S. to SW	30.00 or below and rising slowly	Clearing within a few hours, and fair for several days.
S. to E	29.80 or below and falling rapidly	Severe storm imminent, followed within 24 hours, by clearing, and in winter by colder.
E. to N ..,........	29.80 or below and falling rapidly	Severe northeast gale and heavy precipitation; in winter, heavy snow, followed by a cold wave.
Going to W	29.80 or below and rising rapidly	Clearing and colder.

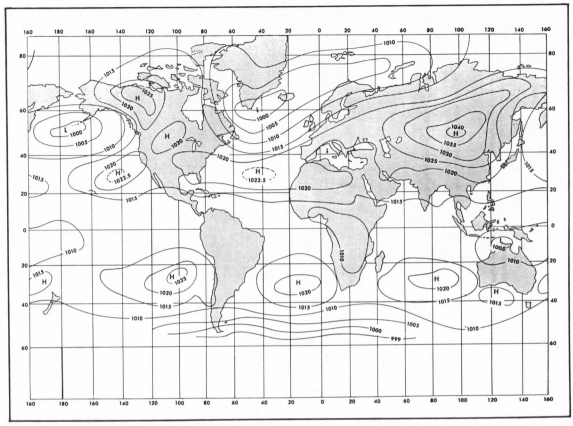

Fig. 8-10. Prevailing world pressure systems (January).

clouds fuse into sheet clouds, and lower or darker clouds develop dark bases.

HOW THE NWS FORECASTS WEATHER

The National Weather Service uses the latest in meteorological and computer technology to prognosticate tomorrow's weather from satellites to climatological data. Let's consider their methods and how they can be applied to your own weather forecasting needs.

The most common method of forecasting weather—for both the amateur and professional meterologist—is known as *syntopic* forecasting. It uses a synopsis or summary of the total weather picture to project into the future. The movement of weather systems is illustrated on a sequence of weather maps, also called *syntopic charts*. Observations noted on these maps are made at thousands of weather stations around the world many times each day (Figs. 8-10, 8-11). Information on the upper atmosphere is also charted daily to assist in understanding and projecting future weather conditions in a given location.

Another method of reading weather, called *statistical* forecasting, uses mathematical equations based on the past behavior of the atmosphere. It says that, since yesterday's temperature was 78, the temperature a year ago was 82, and the normal over the last 12 years for this date is 80, that the high today will be 80 degrees.

Numerical forecasting uses mathematical models based on the physical laws of the Earth's atmosphere using the science of fluid dynamics. Theoretically, complete and precise data on the initial state of the earth's atmosphere, water bodies, and land surfaces, plus a complete understand-

154

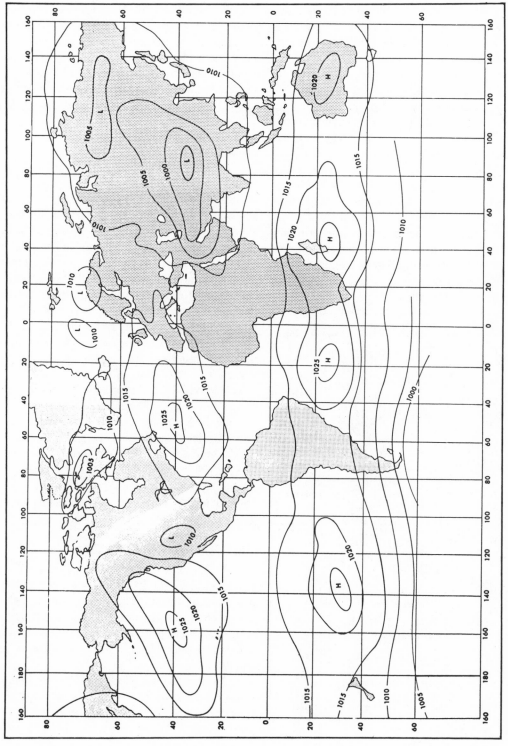

Fig. 8-11. Prevailing world pressure systems (July).

155

ing of the physical laws of heat and moisture transfer could yield near-perfect numerical weather forecasts. Unfortunately, such information isn't fully available.

However, the advent of high-speed computers in the last two decades is changing this situation. Six basic equations expressing the three dimensions of motion and the conservation of heat, moisture and mass are used in numerical models. These equations are solved by computers to offer instantaneous changes at thousands of regularly spaced grid points. Computers at the National Meteorological Center in Suitland, Maryland, compute the weather future in 20 minute intervals for the desired time range of the forecast. This marching forward in time is the essence of numerical prediction used for forecasts up to about five days with a high degree of accuracy. Figure 8-12 shows a typical numerical forecast map.

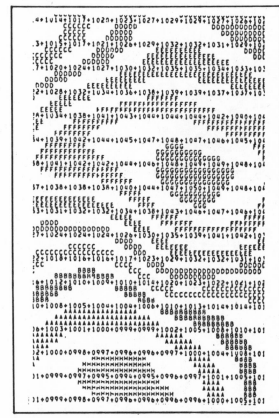

Fig. 8-12. Numerical weather map.

Beyond about 90 days, weather can be predicted just as well through *climatological* forecasting—using the averages of past weather records. This method is also very accurate for *general* weather conditions. After syntopic weather forecasting, climatological forecasting is most often used by the amateur meteorologist for predicting weather.

CLIMATOLOGY

Climatology is the scientific study of climate, the long-term normal weather for a given area or location. The first task of climatology is to describe and thus predict climate by tabulating, statistically analyzing, and graphing or charting records on weather elements and phenomena. The basic elements of weather (temperature, pressure, humidity, precipitation, winds, clouds, sunshine, and visibility) and storms (hurricanes, tornados, and thunderstorms) are observed and recorded. After the observations are used operationally, records of the observations are stored and later used to determine mean (average) extremes, totals, probabilities, or variabilities of climatic elements.

Figures 8-13 through 8-34 show climatic means in the United States for temperature, frost, heating and cooling degree days, precipitation, snowfall, ice, glaze and hail, thunderstorms, fog, winds, humidity, evaporation, sunshine, sky cover, and air pressure. These maps can be extensively useful in predicting weather in your area, planning vacations, and choosing the perfect retirement spot.

CONSIDERING MICRO-CLIMATES

Temperatures may differ considerably from point to point over short distances, particularly in mountainous terrain. These variations usually are due to differences in altitude, slope of land, type of soil, vegetative cover, bodies of water, air drainage, and urban heat effects. These conditions bring about "micro-climates."

An increase in altitude of 1000 feet generally causes a decrease of 3.3°F in the average annual temperature. An extreme case exists in the Grand Canyon area of Arizona between Inner Canyon, ele-

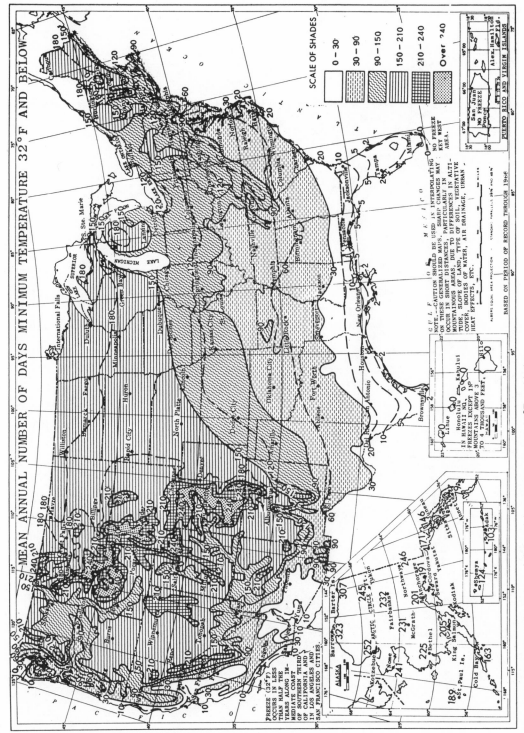

Fig. 8-13. Mean annual number of days minimum temperature 32°F and below.

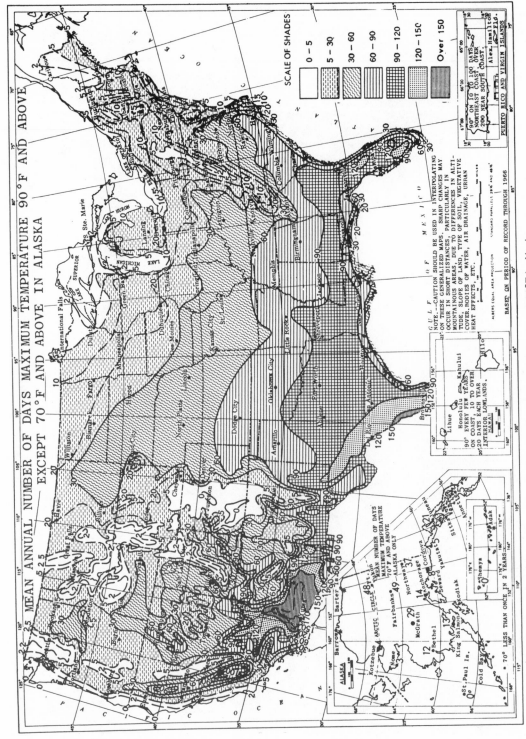

Fig. 8-14. Mean annual number of days maximum temperature 90°F and above, except 70°F in Alaska.

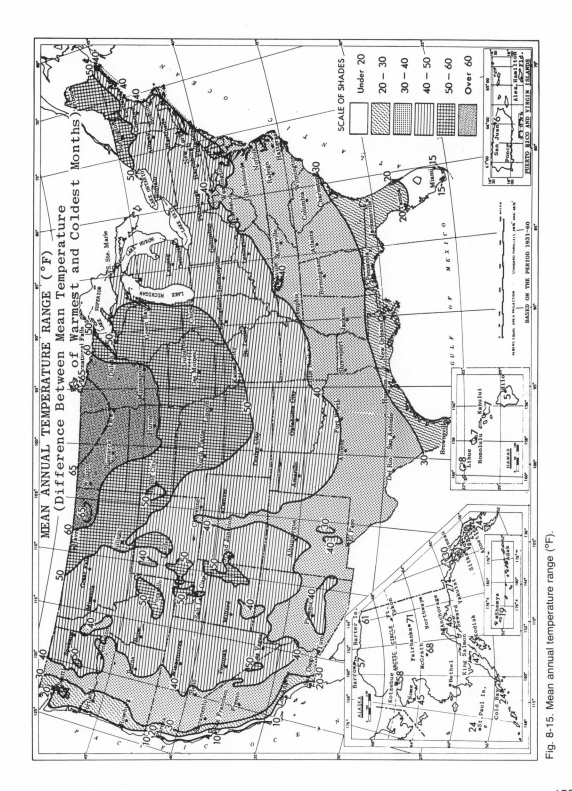

Fig. 8-15. Mean annual temperature range (°F).

159

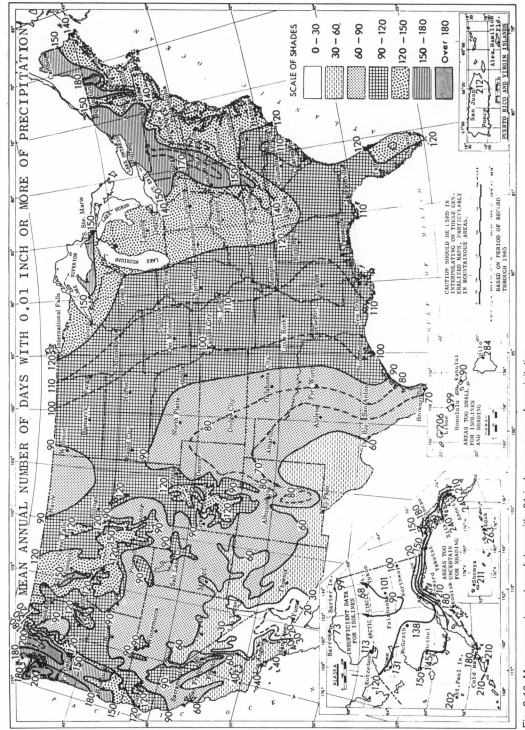

Fig. 8-16. Mean annual number of days with .01 inches or more of precipitation.

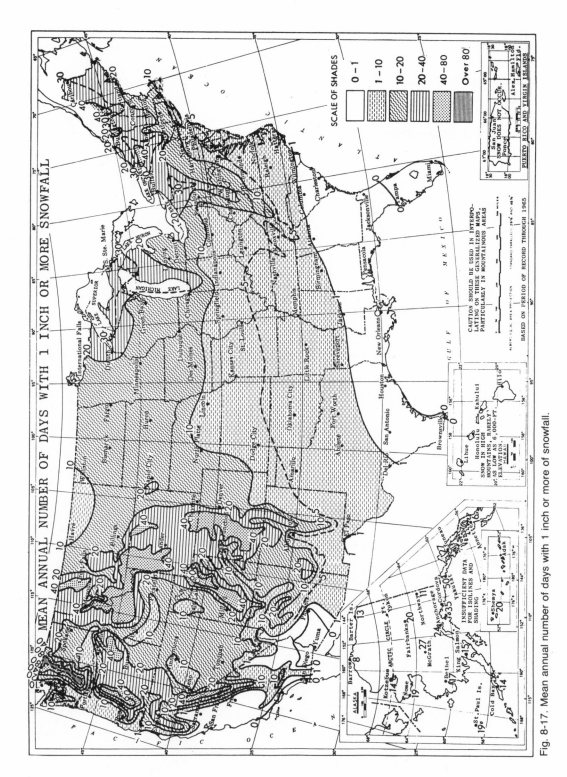

Fig. 8-17. Mean annual number of days with 1 inch or more of snowfall.

161

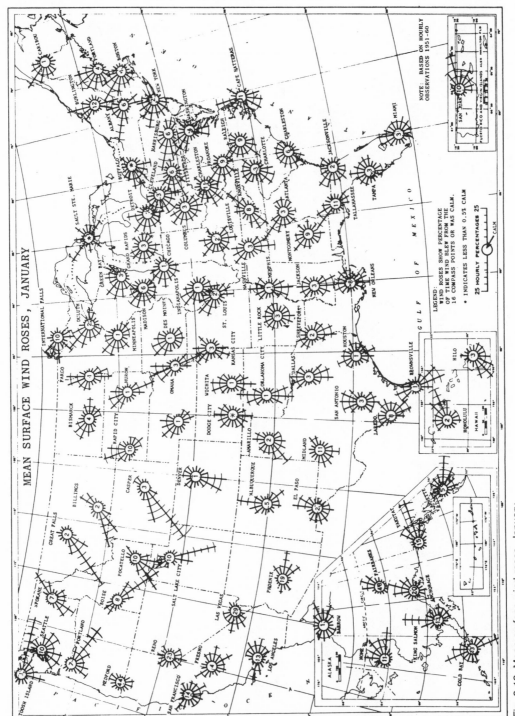

MEAN SURFACE WIND ROSES, JANUARY

NOTE: BASED ON HOURLY
OBSERVATIONS 1951-60

LEGEND:
WIND ROSES SHOW PERCENTAGE
OF TIME WIND BLEW FROM THE
16 COMPASS POINTS OR WAS CALM.
• INDICATES LESS THAN 0.5% CALM
25 HOURLY PERCENTAGES 25

Fig. 8-18. Mean surface wind roses, January.

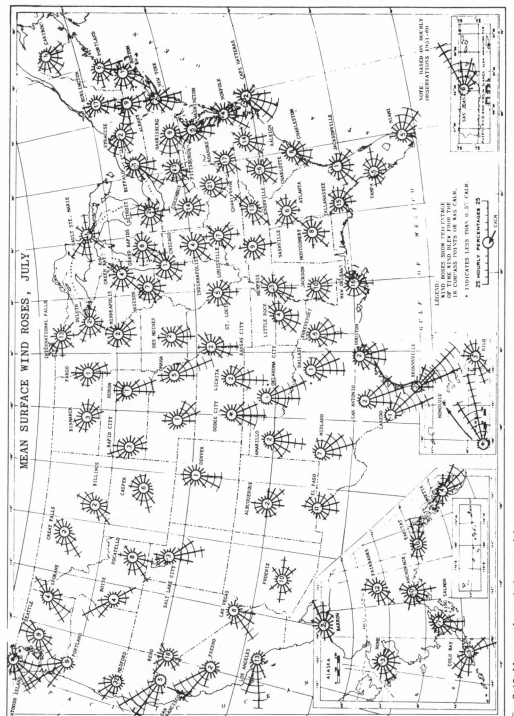

Fig. 8-19. Mean surface wind roses, July.

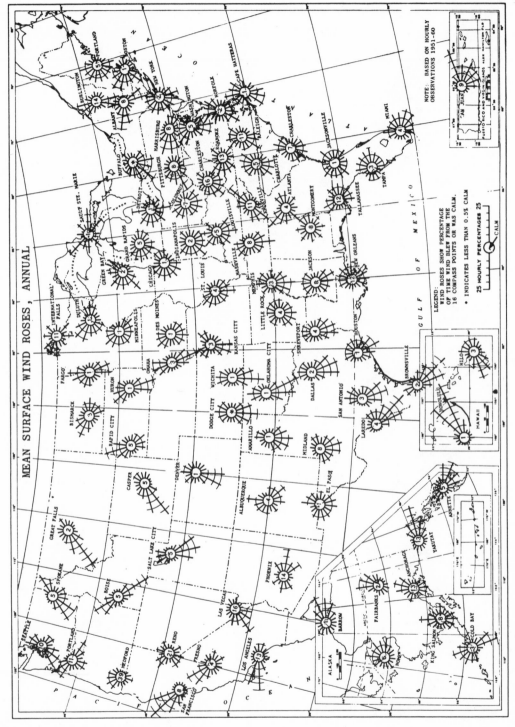

Fig. 8-20. Mean surface wind roses, annual.

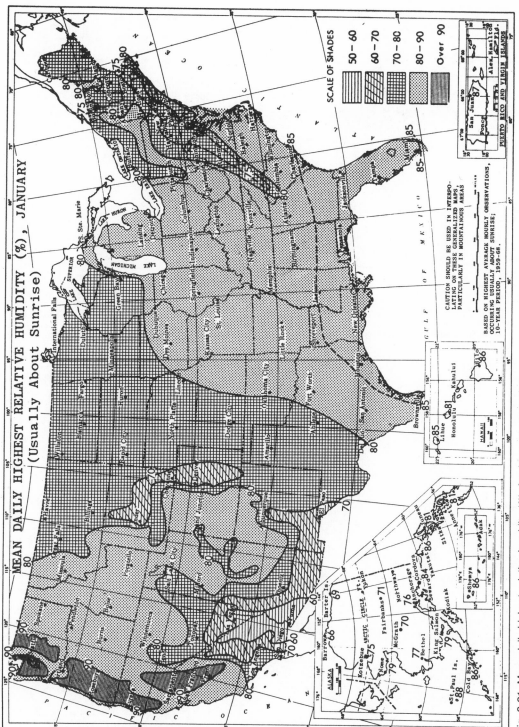

Fig. 8-21. Mean daily highest relative humidity (%), for January, at sunrise.

165

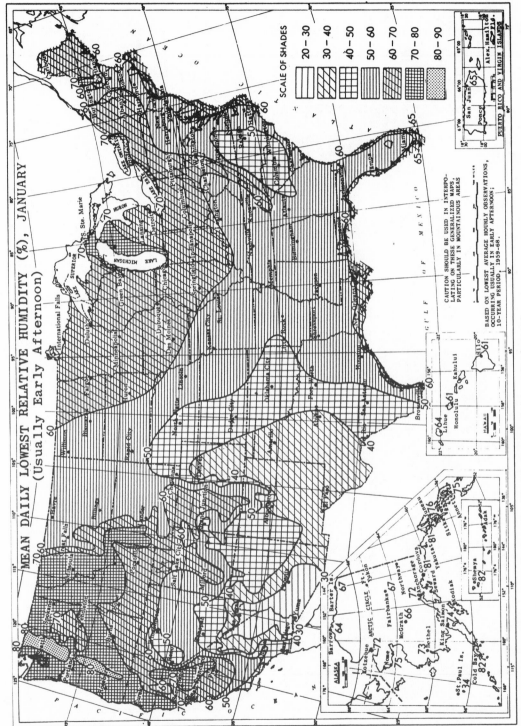

Fig. 8-22. Mean daily lowest relative humidity (%), for January, during early afternoon.

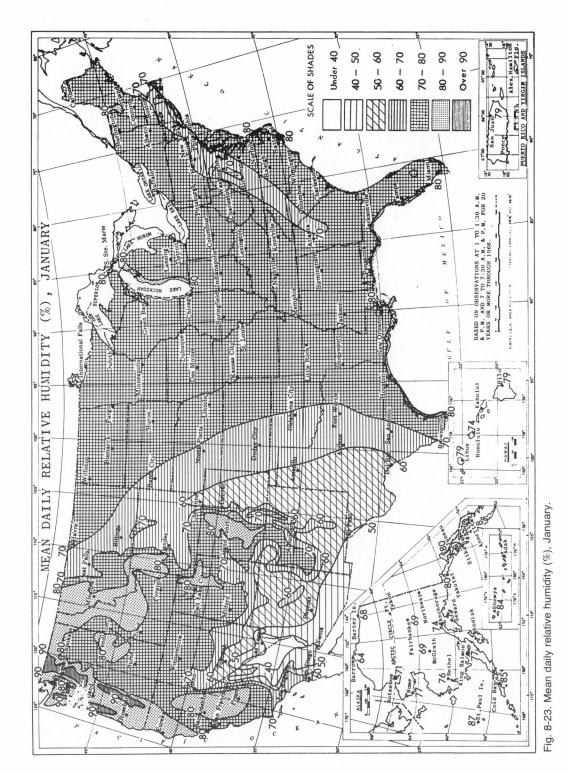

Fig. 8-23. Mean daily relative humidity (%), January.

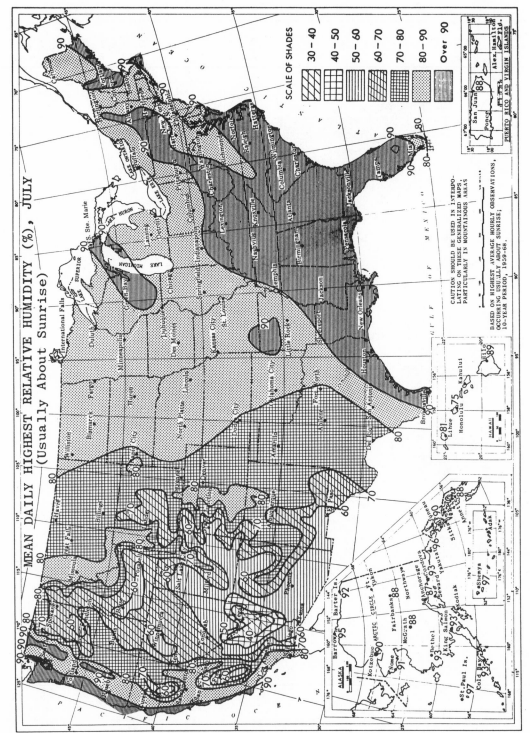

Fig. 8-24. Mean daily highest relative humidity (%), July, at sunrise.

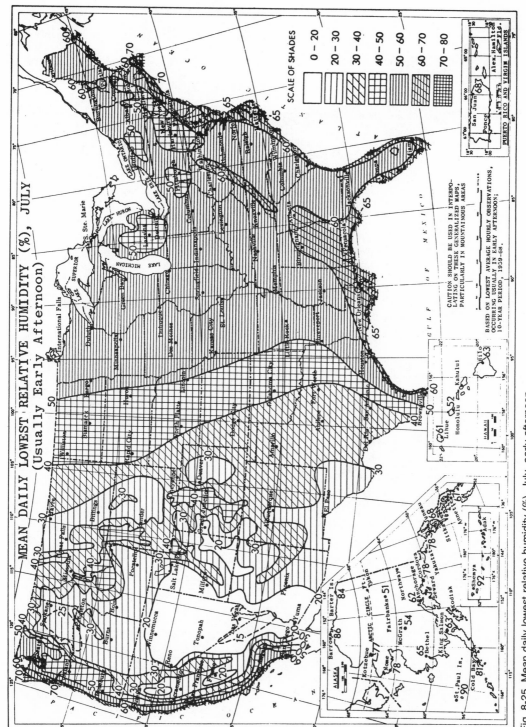

Fig. 8-25. Mean daily lowest relative humidity (%), July, early afternoon.

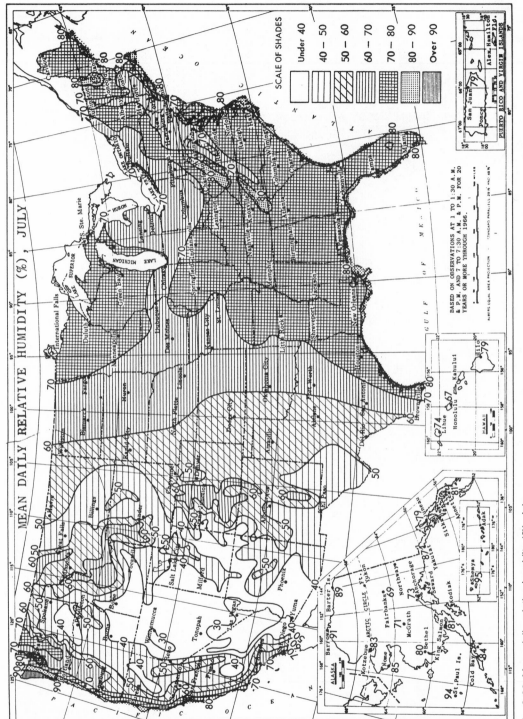

Fig. 8-26. Mean daily relative humidity (%), July.

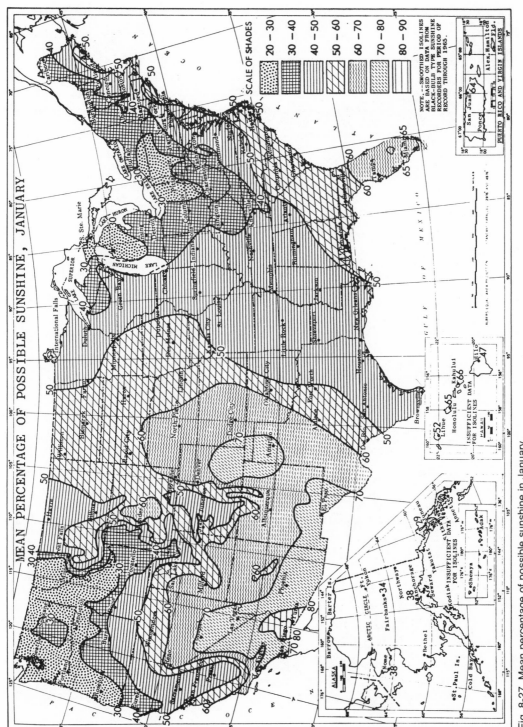

Fig. 8-27. Mean percentage of possible sunshine in January.

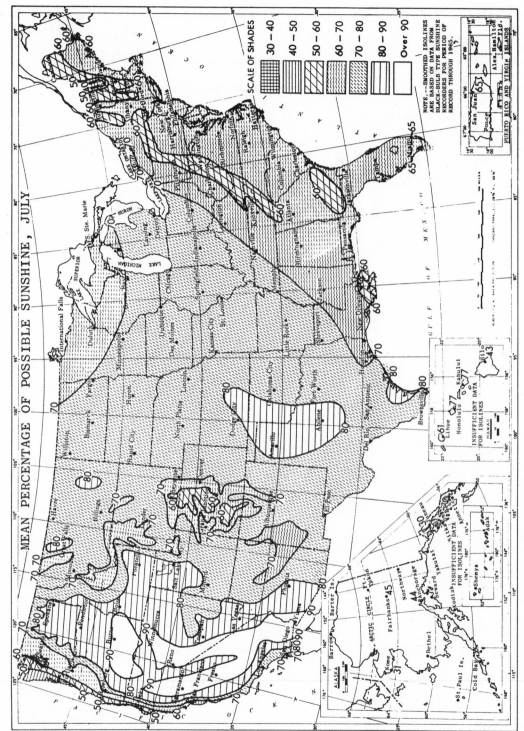

Fig. 8-28. Mean percentage of possible sunshine in July.

172

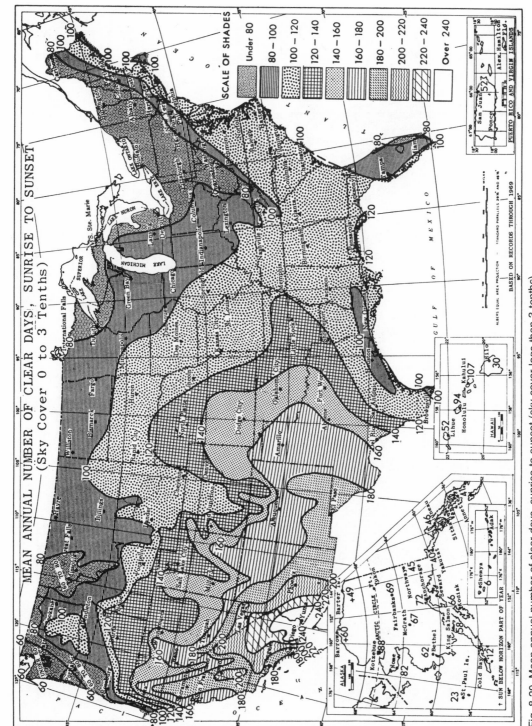

Fig. 8-29. Mean annual number of clear days, sunrise to sunset (sky cover less than 3 tenths).

173

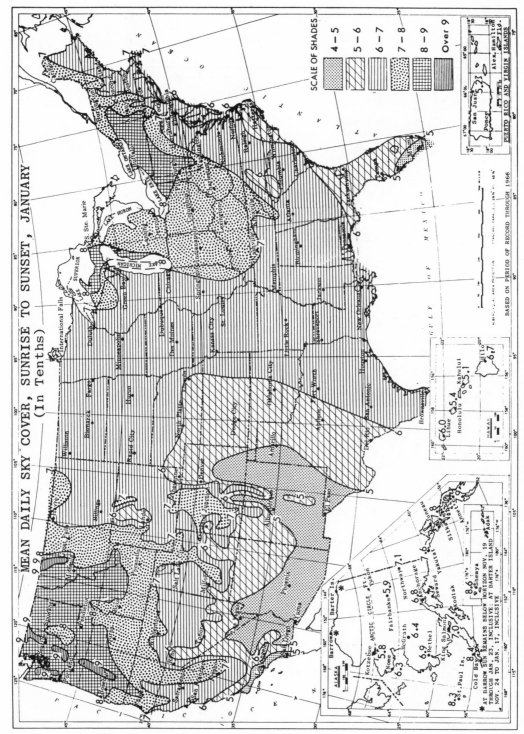

Fig. 8-30. Mean daily sky cover, sunrise to sunset, in January (in tenths).

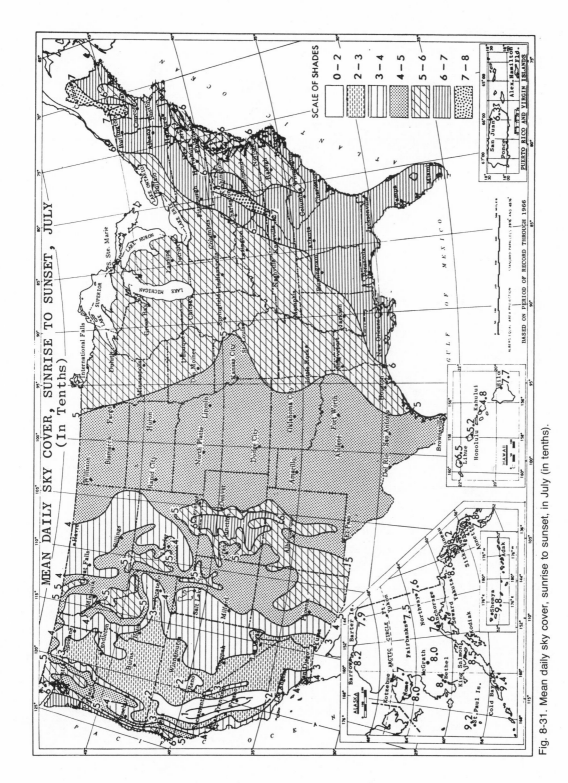

Fig. 8-31. Mean daily sky cover, sunrise to sunset, in July (in tenths).

175

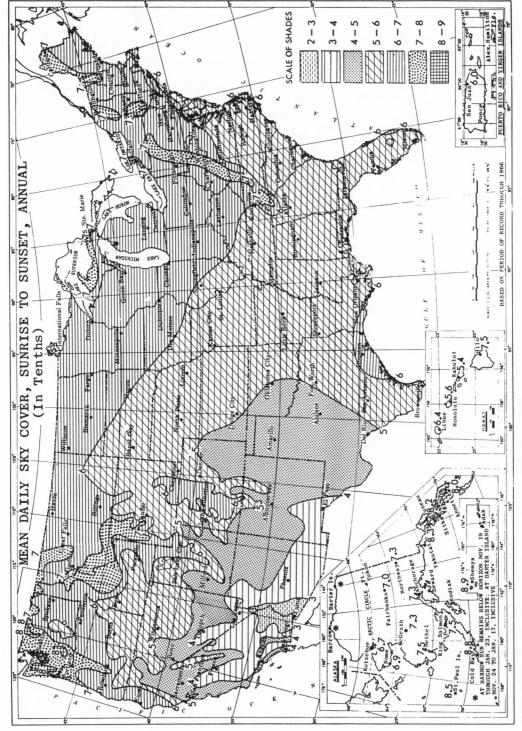

Fig. 8-32. Mean daily sky cover, sunrise to sunset, annually (in tenths).

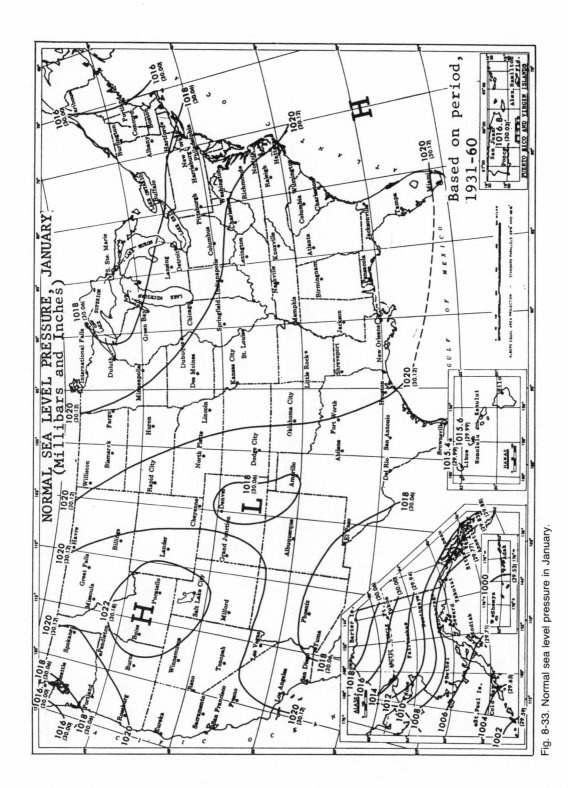

Fig. 8-33. Normal sea level pressure in January.

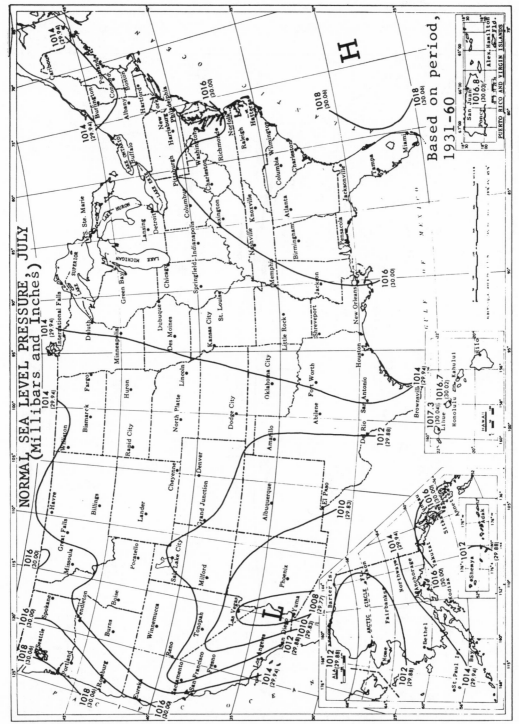

Fig. 8-34. Normal sea level pressure in July.

vation 2500 feet, and Bright Angel, elevation 8400 feet—5900 feet higher, but just five miles away. The normal daily highest and lowest temperatures in July are 106°F and 77°F at the former, and 78°F and 47°F at the latter; in January it's 55°F and 37°F against 37°F and 15°F.

Because of the daily motion of the sun, an eastern slope will be warmer in the mornings and colder in the afternoons than a western slope. A southern slope will remain relatively warm all day, while a northern slope will remain cool.

At night (especially on clear, calm nights) air near the ground cools because of radiative cooling. The cold air is heavier, so if the land is sloping the cold air will drain off and flow to lower elevations. At the original elevation, the cold air will be replaced by warmer air. Farmers use this knowledge by planting hardy vegetation in the bottom of the valleys and more tender vegetables and fruits in the warmer belts just above.

Bare, dry soil will heat and cool more rapidly than heavy, claylike, wet soil, or a soil covered with vegetation. Light colored surfaces will be cooler because they reflect a large portion of the incoming solar radiation, while dark surfaces will be warmer because they absorb much of the solar heat.

Bodies of water, such as the Atlantic and Pacific Oceans, the Gulf of Mexico, the Great Lakes, and to a smaller extent the more than 50,000 small lakes from the eastern Dakotas to Maine, have a moderating effect on adjacent land areas. This effect varies, depending on wind, topography, and size of the water area.

Cities and industrial areas generally average a few degrees warmer and experience less extreme temperatures than the adjacent countryside. This is due to the trapping of solar heat in the city by the complex of buildings, sidewalks, and streets; heat produced in domestic and industrial furnaces, stoves, and incinerators, and the retention of much of this heat by the blanketing effects of smoke, gases, dust, and other pollutants in the air over the city.

The prevailing winds over an area are a strong influence on the type of climate the area will have. Onshore winds over coastal areas usually give such areas a milder and moister climate than if the winds were from the interior of the continent. The prevailing winds on the Pacific coast are from the northwest (Figs. 8-18 through 8-20). Winds over the eastern two-thirds of the conterminous United States are from the northwest or north during January and February, and generally from south to southwest from May through August. In the Texas/Oklahoma area, however, southerly winds prevail from March through December. Easterly winds of the northeast trades prevail over the Florida peninsula except during December and January when the northern winds bring cooler weather.

As you learned earlier, relative humidity is a percentage value that expresses the ratio between the amount of water vapor in the air and the amount of water vapor the air could hold at a given temperature. The higher the temperature, the more water vapor the air can hold. An increase of 20°F usually doubles the capacity of air to hold water vapor, so when air temperature increases 20°F an existing 100 percent relative humidity will decrease to 50 percent. In general, relative humidities are lower during the afternoon when the air temperature reaches its daily maximum, and higher in the early morning when the minimum temperature for the day occurs.

IMPROVING FORECAST ACCURACY

We can all understand the feelings of the irate woman who called up the local weather office early one morning and yelled, "Ya wanna get someone out here to shovel off this six inches of *partly cloudy*?" Weather forecasts are not always accurate. In fact, sometimes they can be 180 degrees off. We all remember rained-on picnics, thunderstorms that never appeared, and wind storms that caught us unprepared.

Unfortunately, meteorology is not an exact science. There are too many elements that play a part in the weather that reaches our location. An unexpected change in a high altitude wind may veer a low pressure system away from us by only 50 miles and give us crystal-clear weather as we pull out umbrellas. Or a distant front may play a slight effect on the local front and modify the path just a few miles—enough to miss you. Or a storm may dissipate before reaching you, passing harmlessly overhead.

However, the major cause of inaccurate weather forecasts is the need to *generalize.* That is, a regional forecast may state that weather will be partly cloudy, but a high range of mountains may condense the clouds as they try to rise above it. The heavier clouds now climb the slope, getting colder and colder until they drop six inches of "partly cloudy" onto a home near the summit. On the other side of the summit, a neighbor complains about the dry weather. These are the micro-climates discussed earlier.

So how can you improve the accuracy of weather forecasts for your micro-climate? First, by recognizing the conditions around you that are not "normal" or typical in your area—mountains in a desert, a large lake near a city, a low valley surrounded by mountains, a large forest nearby.

Then you must consider what affect these elements will have on the weather. Will they increase the temperature? Decrease it? Moderate temperature extremes? Add relative humidity? Reduce winds? Cause tornados or waterspouts? Dissipate thunderstorms? You can do this by reviewing the early chapters of this book or taking a college course in meteorology. Or you can learn by observation and comparison. When the weatherman predicts scattered showers by noon and you don't receive them until nearly two hours later, mark it down in your weather log. Long-term temperature comparisons may indicate that your local temperature is 2°F warmer and 3°F cooler than the "official" air temperatures offered on the weather radio.

Your weather log may also offer information on other differences between your weather and the official weather for your area that is often taken at an airport. Your micro-climate may offer more or less precipitation, faster or slower winds, higher or lower relative humidity, or other elements. A comparison can be useful in modifying the official weather forecast to be more appropriate for your own corner of the world.

By collecting accurate data and interpolating it with the resources of the NWS statellites, computers, and professional meteorologists, you can make highly accurate forecasts of local weather hours, days, and even months in advance. In fact, your forecasts can be even more accurate for your location than those issued by the Weather Service. *You* can be the local "weatherman."

Chapter 9

Using the Weather

It was early on the morning of June 18, 1815. All night long torrential rain, lightning, thunder, and high winds had softened the ground that surrounded and separated the two armies. Rather than attack at dawn, one general decided to wait for the sun to come out and dry the ground so the artillery could be moved forward. The sun never came out that day, but the wait gave the enemy's reinforcements enough time to attack the general from the rear, causing a hasty retreat of Napoleon and his forces at Waterloo. The war was lost.

On the same continent 129 years later, another general worked his climatologists around the clock to predict the ideal conditions for his air, sea, and land forces to overrun the beach at Normandy. General Eisenhower used the weather to his advantage and turned the tide of World War II.

Today, weather forecasts by amateur and professional meteorologists are applied to every aspect of commerce and civilization—from raising crops to flying airplanes to planning picnics to waging war. Weather information can be used to plan department store sales, select horserace winners, estimate factory production, purchase fruits and vegetables, buy and sell grain futures, and many other tasks. Weather permeates our lives whether we understand it or not—and our knowledge of weather can help us appreciate and use the treasures of this world.

AGROMETEOROLOGY

Agrometeorology is a 60 cent word that means using the weather to help in planting, raising, and harvesting food. It's the study of the relationship between weather and climate to agriculture.

Of course, in growing crops, the most important weather element is precipitation. Water is needed in the right amount to ensure adequate moisture for growth—not too much and not too little. Whether you're raising 640 acres of soybeans or a 10-by-10 plot of vegetables, the amount of moisture your "farm" receives is a prime consideration in its location, care, and selection of crops.

Growers depend on both long-term and short-term weather forecasts. They want to know if the growing season will be adequate for sugar beets, but they also need to know if their 80 acres of feeder

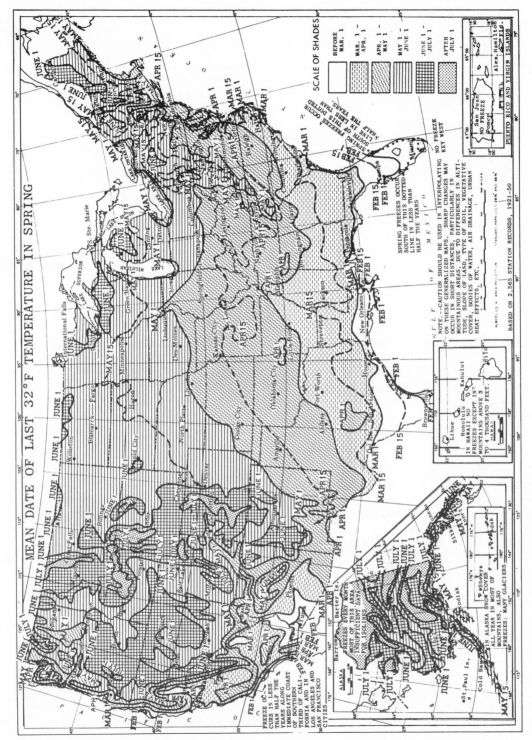

Fig. 9-1. Mean date of last 32°F temperature in spring.

182

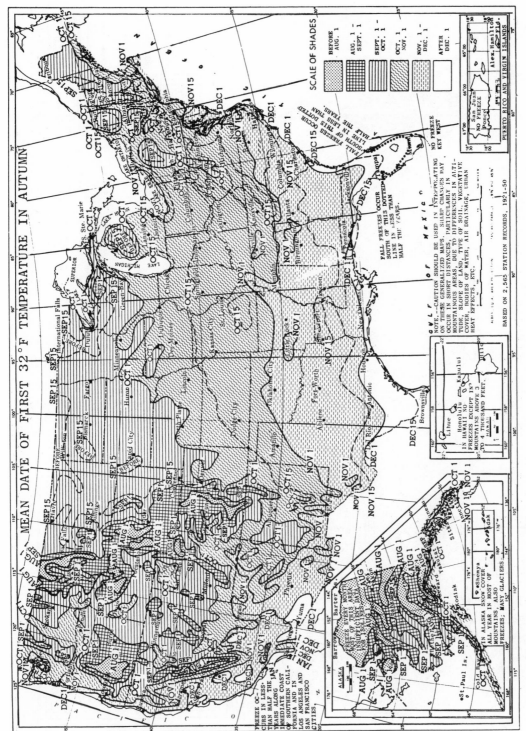

Fig. 9-2. Mean date of first 32°F temperature in autumn.

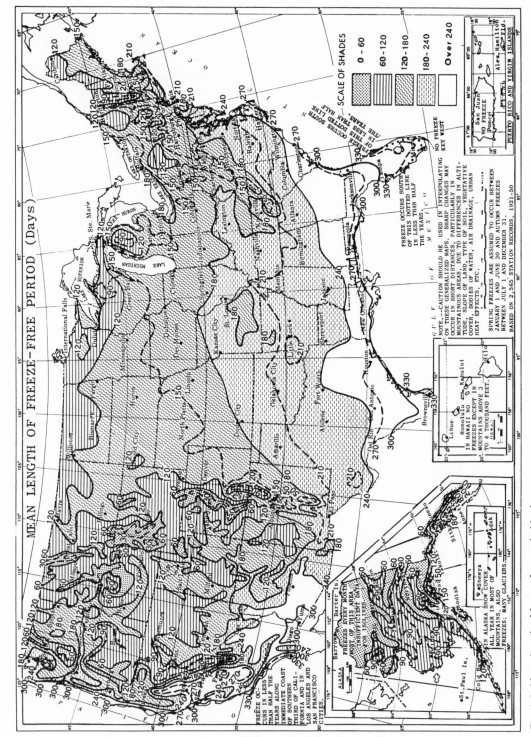

Fig. 9-3. Mean length of freeze-free period in days.

corn are threatened by hail. The key to these problems is in weather records. What do the climatologists say are the growing season or the chance of hail?

By definition, the freeze-free growing period is between the dates of the last 32°F temperature in the spring and the first 32°F temperature in autumn. It is usually referred to as the "growing season" (Figs. 9-1 through 9-3). Temperatures, however, may drop to freezing or below at the ground surface because of radiation on clear, calm nights, even when the offical low temperature measured in the weather shelter is as high as 10°F above freezing. Furthermore, very sensitive plants will be damaged or killed by temperatures as high as 5° to 10°F above freezing. The growth rate of very hardy vegetation, such as winter cereals and many grasses, usually is only restricted by freezing temperatures. Such vegetation remains dormant during persistent cold weather and starts to grow again when moderate warmth returns.

The effect of low temperatures on vegetation depends on the variety and condition of plants be-fore the freeze, the severity and persistence of low temperatures, and protection afforded by a blanket of snow or other covering. Winter wheat, for instance, is economical to grow in the Great Plains as far north as north central South Dakota. It will freeze during many winters if planted farther north. However, winter wheat is grown successfully farther north in the sheltered snow-covered valleys of western Montana. A growing season of at least 180 to 190 days is needed for the growth of cotton, 120 to 130 days for grain corn, and 90 to 100 days for spring wheat. The long hours of summer daylight in northern areas compensate in large measure for the short growing season. For average world temperatures, see Figs. 9-4 and 9-5.

Many other temperature factors determine the location and type of crops the farmer or gardener sows. One such factor is the occasional occurrence of a "hard freeze," usually early in autumn or unseasonably late in spring. Apples, peaches, cherries, and grapes are grown extensively along the Great Lakes, mainly on the leeward (eastern and southeastern) side because of the moderating effects of

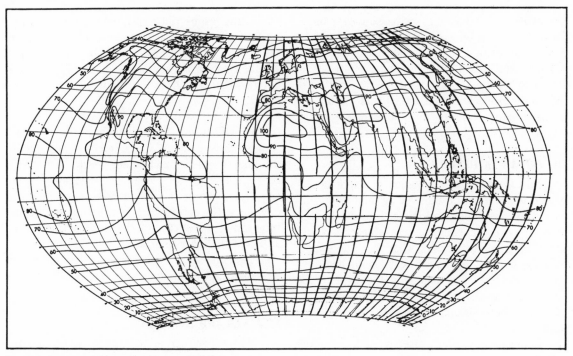

Fig. 9-4. Average July temperature worldwide.

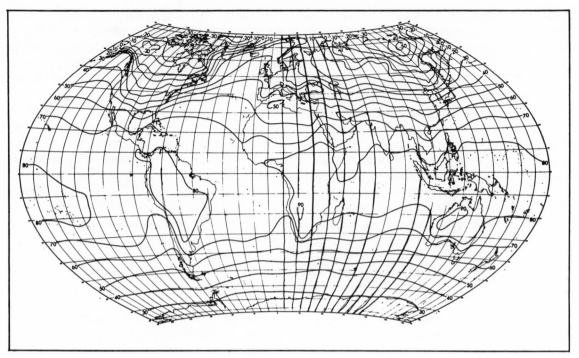

Fig. 9-5. Average January temperature worldwide.

the lake waters on air masses passing over them. This moderating effect retards the premature development of fruits in spring and checks the occurrence of late frosts in spring and early frosts in autumn. On a larger scale, the moderating effect of the Pacific Ocean helps make the Pacific States one of the greatest fruit and vegetable regions of the world.

THE FARM WEATHERMAN

In this age of specialization, it might be expected that the big business of agriculture has its own specialized meteorologists. It does. In fact, there are many men and women who make a full-time living telling farmers what they can expect over the horizon tomorrow, next month, and even next year. They can also help the commercial farmer decide when and how to sell his crop, because weather also plays a major part in food marketing.

Some farmers work directly with these agrometeorologists to plan crops and marketing, while others subscribe to printed and recorded services of a more general nature. Still others serve as their own weather forecasters by setting up a weather station and keeping a weather log, or combining weather entries into their farm operations log.

The home gardener can improve crop yield and rotation with a simple understanding of weather and weather forecasting. With the help of a garden shop, the home gardener can select "early" and "late" crops that fit the local and micro-climate growing season for the most efficient use of the land and elements. The gardener can also estimate watering requirements in advance to decide which food crops will cost the least to grow.

In many cases, the gardener or farmer can get help on local weather and crops through the County Extension Service.

WEATHER AND THE PILOT

The second largest user of weather information is the private pilot. Commercial aircraft often fly well above the weather and only need to consider winds aloft and the weather at the points of departure and destination, but the pilot of the small airplane is constantly surrounded by the weather. The wind governs his true airspeed. The visibility and cloud cover dictates his altitude and even whether or not he can fly. In spite of the excellent meteorological services available to general aviation pilots, the weather remains a major cause of accidents.

The major concerns the pilot must have about weather are lack of visibility, turbulence, icing, hail, and strong surface winds. Singly, or together in storms, these are the impediments to safe flight.

The visibility problem may be caused by one of several kinds of fog, clouds, storms, snow, haze, smog, and dust.

The most dangerous turbulence is found in or near thunderstorms, low altitude wind shears, mountains, and in the wakes of the large aircraft.

Aircraft icing may be classified into two main groups: *structural* and *powerplant*. Structural icing may be anticipated anytime the free temperature is 0°C or colder and the humidity high (narrow spread between dewpoint and temperature at your altitude). Carburetor ice may form in conditions of high humidity with temperatures as low as 10°C and as high as 25°C. It is most serious when the temperature and dewpoint approach 20°C. Air intake ducts are most likely to ice when the temperature and dewpoint are near 10°C or less.

The greatest danger from hail outside of a thunderstorm, is directly beneath a thunderstorm, and in clear air beneath the anvil top of a mature and dissipating thunderstorm that tops 35,000 feet or more.

The dangers generated by strong surface winds are mostly to taxiing, landing, and takeoff operations. Usually, these are purely local winds that create turbulence downwind from surface protuberances.

Dangerous low-altitude wind shear near airports is caused by a temperature inversion in which a layer of fast-moving warm air is above colder, relatively calm air next to the surface.

Let's take a closer look at each weather ele-

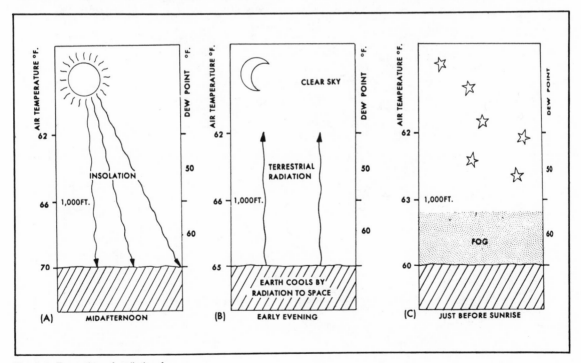

Fig. 9-6. Formation of radiation fog.

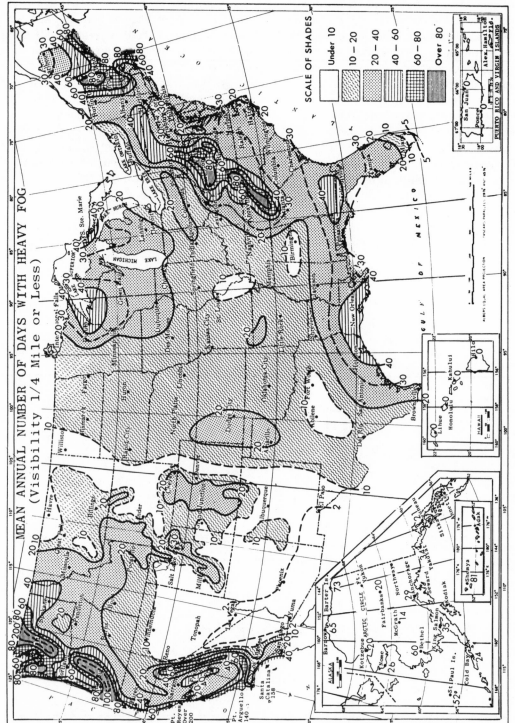

Fig. 9-7. Mean annual days with heavy fog (visibility of ¼ mile or less).

ment and how it aids or impedes the private pilot. Strap on your leather cap, adjust your goggles, and make one more wrap with your neck scarf.

Fog and Flying

Fog may form either as the result of moisture-laden air cooling to its dewpoint, or by the addition of moisture to air near the ground (Fig. 9-6). Ideal conditions for the formation of fog are high relative humidity (small spread between the temperature and dewpoint), plenty of condensation nuclei (tiny particles from sea spray, dust, or products of combustion), light surface winds, and a cooling process that starts the condensation. Therefore, fog is more prevalent over coastal areas where moisture and nuclei are plentiful. However, fog often occurs in some industrial areas inland, even when humidity is less than 100 percent, due to a plethora of condensation nuclei (Fig. 9-7).

Fog is more frequent in the colder months and fog which forms at 15°C or lower may be made up of ice crystals of super-cooled water droplets. Aircraft flying through the latter may be subject to sudden structural icing because the impact of the plane is all that is needed to transform the super-cooled droplets into instant ice, with the airplane's surfaces providing the necessary nuclei.

Technically, ground fog is radiation fog and it forms on clear, calm nights when the ground cools the air in contact with it to the dewpoint temperature. It usually is shallow, favors relatively level ground, and generally burns off well before noon. The fact that it is known to be shallow can boobytrap the unwary, however, because advection fog, which is deeper, can form above it.

Advection fog is a coastal fog that forms when moisture air moves over colder ground or water. It can form concurrently with ground fog, but is usually deeper than ground fog. When the surface wind is greater than 15 knots, it will lift to form a layer of low stratus clouds. (See Table 9-1.)

Upslope fog results when moist, stable air is cooled by forced ascension up a sloping land surface. It is common over the eastern slopes of the Rockies and is found less frequently on the eastern side of the Appalachians.

Table 9-1. Reportable Visibility Value (Miles).

Increments of Separation (Miles)					
1/16	⅛	¼	½	1	5
0	⅜ 1¼	2	2½	3 10	15
1/16	½ 1⅜	2¼	3	4 11	20
⅛	⅝ 1½	2½		5 12	25
3/16	¾ 1⅝			6 13	30
¼	⅞ 1¾			7 14	35
5/16	1 1⅞			8 15	40
⅜	1⅛ 2			9	etc.

Steam fog forms over water when the movement of cold air over much warmer water sets up intense evaporation. Because steam fog, unlike advection fog, forms over a warm surface, the heating from below tends to make the air unstable. Therefore, turbulence and icing may be expected with this kind of fog. It is mostly observed in the polar regions, but is sometimes found over lakes and rivers in the northern United States during autumn.

Precipitation-induced fog is caused by the addition of moisture to the air through evaporation of rain or drizzle. It is most frequently associated with warm fronts, though it can occur independently. When associated with a front, this kind of fog usually forms very quickly and covers a large area.

Chasing Clouds

Visible evidence of what the atmosphere is doing, its movement, water content, and stability, can be read in the clouds.

In stable air, stratus clouds form in layers with little or no vertical development. Thickness may range from a few hundred feet to several thousand. Precipitation is drizzle, continuous light rain, or snow. There's little or no turbulence associated with stratus formations, but the danger of structural icing is great if the free air temperature is 0°C or colder. Carburetor ice is likely if temperature is 25°C or lower. In short, the presence of middle and/or low stratus clouds usually means poor flying weather.

Cumulonimbus clouds are also called "thunderheads." They are rarely larger than ten miles in diameter, but often run in packs, passing out disas-

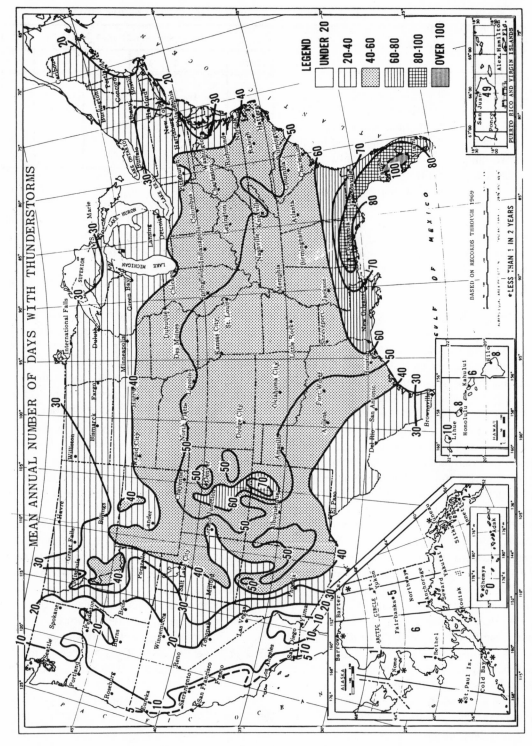

Fig. 9-8. Mean annual number of days with thunderstorms.

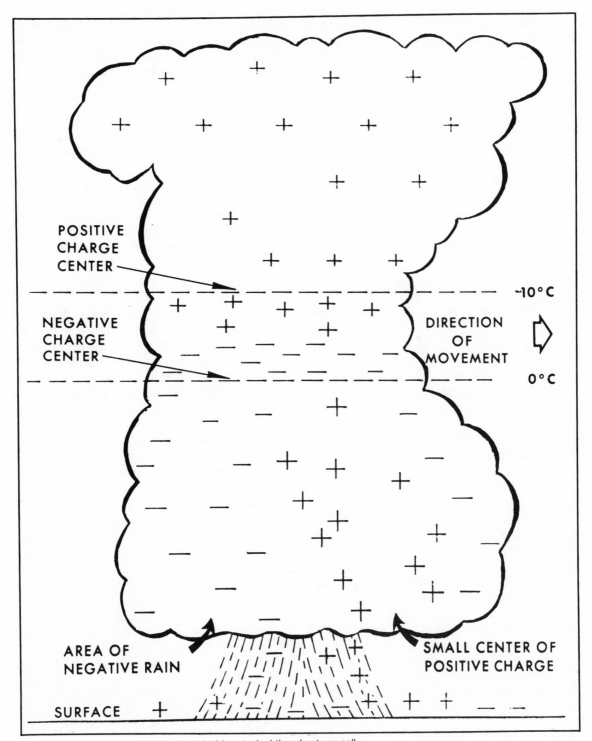

Fig. 9-9. Location of electrical charges inside a typical thunderstorm cell.

terous weather for a hundred miles and lasting up to eight hours (Fig. 9-8). And since they often reach from ground level to as high as 60,000 feet, the pilot can't go under, over, or around them. The only things to do is sit them out on the ground. Figure 9-9 shows the development of these powerful cloud systems.

Squall lines often develop 50 to 300 miles ahead of fast-moving cold fronts, bringing severe weather such as heavy hail, destructive winds, tornados, and thunderstorms.

Turbulence

The four main causes of air turbulence that make things rough on the pilot are vertically moving air in convective currents, air moving around or over mountains and other obstructions, wind shear, and the passage of large aircraft.

If you've flown in small aircraft you are familiar with convective currents that cause bumpiness in lower altitudes during warm weather. They are local in nature, with both ascending and descending currents. When sufficient moisture is present, this process will produce cumulus clouds.

Turbulence that results from air near the surface flowing over rough terrain, trees, buildings, and hills may set up tricky situations for lightplanes during landing and takeoff operations. When the wind is light, such turbulence is minimal; but in wind speeds above 20 knots, the flow may be broken up into irregular eddies that can persist downwind to create a hazard in the landing area.

Especially dangerous are the winds that blow up the windward slope of mountains. The air above them is usually fairly smooth in a stable atmosphere, however, as wind spills down the leeward side it produces downdrafts and turbulence. If winds are strong, you need at least 2000 feet between you and the ridge for a reasonably safe crossing.

Either vertical or horizontal, wind shear is a localized condition that occurs when a stream or mass of air moves at a relatively high velocity and usually in a direction different than the air mass directly adjacent to it. This condition is fairly common over the northern U.S. and in Canada during winter where temperature inversions occur near the surface and the terrain below—usually a valley—holds cold, calm air beneath a moving layer of warm air.

Finally, wake turbulence is, in effect, a pair of tiny twin tornados generated by the airfoils of large aircraft. These twisting, turbulent columns of air stream behind the big planes from each wingtip with a downward force of about 1500 feet per minute and impose a roll-rate of up to 80 degrees per second on a light aircraft caught within one. Normally, wake turbulence is encountered only within a minute or two after the passage of a big transport craft.

Aircraft Icing

Aircraft icing is a major weather hazard to aviation. Its formation can dangerously distort airfoil shapes, add drag and weight, and induce structural vibration. Ice deposits of only one-half inch on the leading edge of airfoils (wings, for you nonpilots) on some aircraft reduce their lifting power as much as 50 percent, increase drag by an equal amount, and greatly increase the stall speed (at which a plane no longer can fly).

Clear ice forms on structural parts of the aircraft in the shape of a blunt nose with gradual tapering toward its trailing edges. It is the most serious form because it adheres firmly and is difficult to remove. Rime ice forms by instantaneous freezing of small super-cooled droplets. This traps a large amount of air, giving the rime its opaque, milky appearance. It has little tendency to spread and is comparatively easy to remove if the aircraft is equipped with deicing systems. Frost can be troublesome in flight when a cold aircraft descends from a zone of subzero temperatures through a zone of warmer air with high relative humidity. Windshields are especially susceptible to frost under such conditions. Frost collecting on the upper surfaces of an aircraft parked outside overnight is usually thin and doesn't look as if it could affect the lift and drag of the plane very much. Don't be deceived; it *can*. *Any* frost is too much frost during takeoff.

WEATHER AND BOATERS

Boating is a growing pastime in the United States and many new boaters discover the waters of our country each year. They, especially, must become aware of the weather and the part it plays in safe boating.

The boater should understand the basics of weather as outlined earlier in this book as well as specific knowledge on marine weather.

Figure 9-10 shows marine weather service charts available throughout the United States offering National Weather Service office telephone numbers and locations of warning display stations. These charts are available at local marinas and marine chart dealers, or by ordering from Distribution Division (C44), National Ocean Survey, 6501 Lafayette Avenue, Riverdale, MD 20840.

Figure 9-11 shows the common display signals used by marine weather stations to indicate sea conditions. They are called Small Craft Advisories and cover a wide range of wind and sea conditions for boats of many sizes and designs. Hurricane warning signals are also offered.

The first step in safe boating is obtaining the latest available weather forecast for your boating area. Where they can be received, the NOAA Weather Radio continuous broadcasts mentioned earlier in the book are the best way to keep informed of expected weather and sea conditions. If you hear on the radio that warnings are in effect, or see flags or lights at warning display stations, don't venture out on the water unless you are confident that your boat can be navigated safely under forecast conditions of wind and sea.

Once you're out on the water, keep a weather eye out for the approach of dark, threatening clouds that may foretell a squall or thunderstorm, any steady increase in wind or sea, or an increase in wind velocity opposite in direction to a strong tidal current. A dangerous rip tide condition may form steep waves capable of broaching a boat.

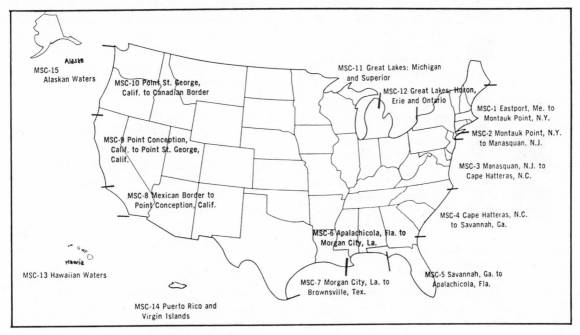

Fig. 9-10. Location of marine weather service charts.

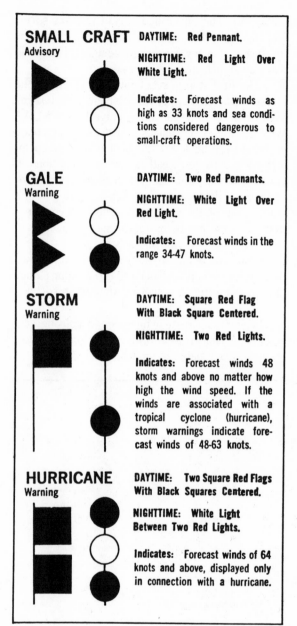

SMALL CRAFT Advisory	**DAYTIME:** Red Pennant.	
	NIGHTTIME: Red Light Over White Light.	
	Indicates: Forecast winds as high as 33 knots and sea conditions considered dangerous to small-craft operations.	

GALE Warning	**DAYTIME:** Two Red Pennants.	
	NIGHTTIME: White Light Over Red Light.	
	Indicates: Forecast winds in the range 34-47 knots.	

STORM Warning	**DAYTIME:** Square Red Flag With Black Square Centered.	
	NIGHTTIME: Two Red Lights.	
	Indicates: Forecast winds 48 knots and above no matter how high the wind speed. If the winds are associated with a tropical cyclone (hurricane), storm warnings indicate forecast winds of 48-63 knots.	

HURRICANE Warning	**DAYTIME:** Two Square Red Flags With Black Squares Centered.	
	NIGHTTIME: White Light Between Two Red Lights.	
	Indicates: Forecast winds of 64 knots and above, displayed only in connection with a hurricane.	

Fig. 9-11. Marine weather advisory signals.

Of course, check radio weather broadcasts for the latest forecasts and warnings. Heavy static on your AM radio may be an indication of a nearby thunderstorm. If a thunderstorm catches you while afloat, you should remember that not only gusty winds but also lightning pose a threat to safety. The best rules are to stay below deck if possible, keep away from metal objects that are not grounded to the boat's protection system, and don't touch more than one grounded object at the same time or you may become a shortcut for electrical surges passing through the protection system.

A bimonthly publication produced by the National Weather Service, *The Mariner's Weather Log,* offers articles on marine meteorology and climate, available through the Superintendent of Documents, U.S. Government Printing Office, Washington, DC 20402.

INDUSTRIAL METEOROLOGY

Commerce and industry also depend on the weather and can use it to their advantage in production and trade. Weather influences transportation by truck, train, barge, ship, and plane. Weather, in the form of heavy storms, can also reduce or completely stop production through the loss of the labor force for a few days or weeks. Rain, snow, and high humidity can cause rapid deterioration of raw and finished products left out in the open. Weather can also play an important part in how wastes are discharged from a factory. Inversion layers may not allow wastes to be put into the air. A lack of snow run-off can reduce river levels and force industries to find alternative and more expensive methods of discharging wastes.

Even safety is relative to weather. In coal mines, gas flows out of the coal when the barometric pressure lowers. This increases the chances of explosions within the mine. And, as you learned in Chapter 4, the weather can change human personalities for the better or worse, changing productivity as well as safety.

WEATHER AND COMFORT

One of the most practical of weather statistics is the *heating degree day*—the number of degrees the average temperature for the day is below 65°F. For example, a day with an average temperature of 50°F has 15 heating degree days, while a day with an average temperature of 65°F or higher has no heating degree days. The amount of heat required at lower temperatures is proportional to the degree-

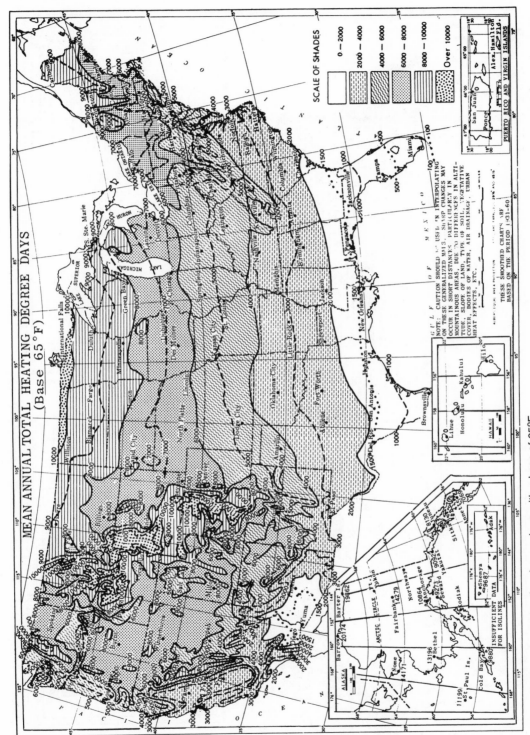

Fig. 9-12. Mean annual heating degree days, with a base of 65°F.

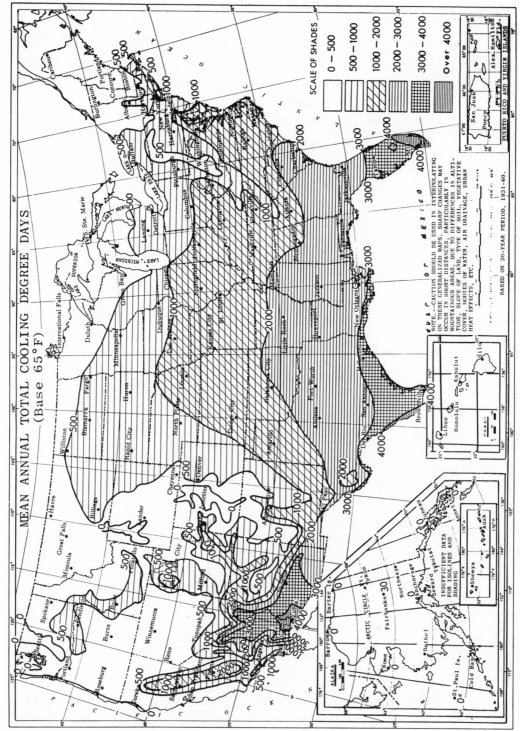

Fig. 9-13. Mean annual total cooling degree days, with a base of 65°F.

196

day value. A fuel bill usually would be about twice as high for a month with 1000 heating degree days as for a month with 500 heating degree days.

In contrast, cooling degree days are the number of degrees the average temperature for the day is above 65°F. Figures 9-12 and 9-13 show heating and cooling degree day averages in the United States.

More important to human comfort is the *Effective Temperature*. The Effective Temperature is an index of the degree of warmth experienced by the body on exposure to different combinations of temperature, humidity, and air movement. This measurement was developed in a series of studies in 1923 at the Research Laboratory of the American Society of Heating and Air Conditioning Engineers. Figure 9-14 shows the relation of effective temperature to comfort.

The Temperature-Humidity Index is a measurement developed in 1958 as a modification of the Effective Temperature method. The Temperature-Humidity Index (THI), originally called the Discomfort Index, can be computed from one of the following equations:

$$THI = 0.4 \, (td + tw) + 15$$
$$THI = 0.55td + 0.2tdp + 17.5$$
$$THI = td - [(0.55 - 0.55RH) \, (td - 58)]$$

where

td = dry bulb temperature
tw = wet bulb temperature
tdp = dewpoint temperature
RH = relative humidity

All "t" values are in degrees Fahrenheit and the "t" values and the RH in each equation are simultaneous readings. The relative humidity is used in its decimal form (i.e., 0.65 instead of 65 percent).

A temperature-humidity index of 60 is generally used as a base for comparison. In summer, relatively few people will be uncomfortable because of heat and humidity while the index is 70 or below, about half the population will be uncomfortable when the index reaches 75, and almost everyone will be uncomfortable when it reaches 80.

WIND CHILL

The effect of a given temperature on the human body is greatly modified by wind and humidity conditions. The effects of humidity were shown above in the section on Effective Temperature. The effect of wind is to lower body temperature by evaporating perspiration or merely by advecting (blowing) heat away from the surface of the skin. Table 9-2 shows the temperature you feel on your skin under given conditions of temperature and wind. Note that the chilling effect increases as the wind speed increases. When winds exceed 40 miles per hour, there is little additional chilling effect.

USING DAILY WEATHER

The conditions of weather can be used in everyone's life, whether you are a farmer, gardener, flier, traveller, industrial plant operator, retail store manager, sports figure or fan, home executive (*nee* "housewife"), student, teacher, or truck driver.

Weather can be applied through understanding the basics of how and why it forms in the atmosphere into air masses and fronts. We must also understand those elements that directly reach us, such as wind, pressure, temperature, humidity, precipitation, and clouds.

Weather is used by watching its patterns and using these rules to predict forthcoming weather. The signs come from observation of the sky, nature, and ourselves. These observations are recorded in diaries and logs to retain them for understanding weather trends and typical weather conditions throughout the year. Weather elements can be recorded using the simplest instruments you can make or find yourself, or by looking over the shoulder of those with sophisticated weather measuring devices—such as the National Weather Service and private meteorologists.

Weather data is then collected from its many sources and interpreted as "forecasts" through one or more popular methods, depending on the equipment and expertise available. Yet even the person with no weather instruments can often predict the

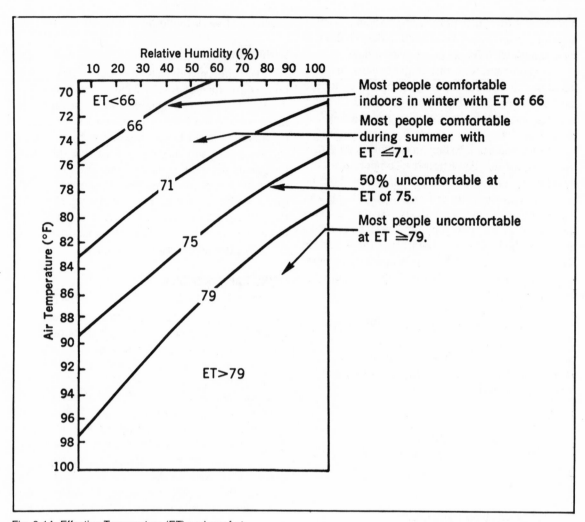

Fig. 9-14. Effective Temperature (ET) and comfort.

Table 9-2. Wind Chill Equivalent Temperature Chart.

								Dry bulb temperature (°F)												
		45	40	35	30	25	20	15	10	5	0	−5	−10	−15	−20	−25	−30	−35	−40	−45
	4	45	40	35	30	25	20	15	10	5	0	−5	−10	−15	−20	−25	−30	−35	−40	−45
	5	43	37	32	27	22	16	11	6	0	−5	−10	−15	−21	−26	−31	−36	−42	−47	−52
	10	34	26	22	16	10	3	−3	−9	−15	−22	−27	−34	−40	−46	−52	−58	−64	−71	−77
	15	29	23	16	9	2	−5	−11	−18	−25	−31	−38	−45	−51	−58	−65	−72	−78	−85	−92
Wind speed (mi/h)	20	26	19	12	4	−3	−10	−17	−24	−31	−39	−46	−53	−60	−67	−74	−81	−88	−95	−103
	25	23	16	8	1	−7	−15	−22	−29	−36	−44	−51	−59	−66	−74	−81	−88	−96	−103	−110
	30	21	13	6	−2	−10	−18	−25	−33	−41	−49	−56	−64	−71	−79	−86	−93	−101	−109	−116
	35	20	12	4	−4	−12	−20	−27	−35	−43	−52	−58	−67	−74	−82	−89	−97	−105	−113	−120
	40	19	11	3	−5	−13	−21	−29	−37	−45	−53	−60	−69	−76	−84	−92	−100	−107	−115	−123
	45	18	10	2	−6	−14	−22	−30	−38	−46	−54	−62	−70	−78	−85	−93	−102	−109	−117	−125

weather in his or her area more accurately than the weatherman's Area Forecasts by observation and knowledgeable interpretation. All the amateur meteorologist needs is a weather radio, a simple thermometer or two, and an understanding of why and how weather works in our lives. This knowledge can then be applied to decide what clothing to wear, plan vacations, predict when business will increase or decrease, decide on appropriate outdoor activities throughout the year, anticipate and prepare for adverse weather, predict the moods and energy level of people, and take advantage of various weather conditions.

Most important, learning how weather affects our daily lives can give each of us a greater appreciation for the earth on which we live and its Creator.

Appendix
Weather Equipment Suppliers

To help you in setting up your own weather station, here is a list of weather instrument and equipment suppliers and manufacturers by location.

California
Abbeon Cal, Inc.
123-13 Gray Avenue
Santa Barbara, CA 93101
Hygrometers, thermometers, anemometers

Echo Science Corp.
485 East Middlefield Road
Mountain View, CA

Hy-Cal Engineering
12105 Los Nietos Road
Santa Fe Springs, CA 90670

Kahlsico International Corp.
P.O. Box 947
San Diego, CA

Kurz Instruments, Inc.
P.O. Drawer 849
Carmel Valley, CA 93924

LST Electronics, Inc.
1825 Eastshore Highway
Berkeley, CA

Meteorology Research, Inc.
P.O. Box 637
Altadena, CA 91001
Instruments

Oceanographic-Meteorological Instrument Co., Inc.
737 West Main Street
El Cajon, CA
All types, including agrometeorology

Sierra-Misco, Inc.
1826 Eastshore Highway
Berkeley, CA

Silver Instruments, Inc.
2346 Stanwell Drive
Concord, CA

Systron-Donner Corp.
1 Systron Drive
Concord, CA

Templine, Inc.
23555 Telo Boulevard
Torrance, CA

Thermonetics Corp.
P.O. Box 9112
San Diego, CA
*Hot wire anemometers,
transducers, heat flow*

**Weather & Wind Instrument
& Equipment Co.**
734 East Hyde Park Boulevard
Inglewood, CA 90302

**Weather Measure Division
of Systron-Donner**
P.O. Drawer 41257
Sacramento, CA 95841

Weathertronics
P.O. Box 1437
West Sacramento, CA 95691

Westberg Manufacturing, Inc.
3402 Westach Way
Sonoma, CA
Anemometers, wind vanes

Western Fire Equipment, Co.
440 Valley Drive
Brisbane, CA
Psychrometers, rain gauges

Connecticut
Airflo Instrument Co.
50 Addison Road
Glastonbury, CT
Wind measuring instruments.

Indecor, Inc.
9 Depot Street
Milford, CT
*Wall and table weather
instruments.*

Kenneth Lynch & Sons
180 Danbury Road
Wilton, CT 06897
Weather vanes.

Florida
Atkins Technical Inc.
3301 SW 40th Boulevard
Gainesville, FL
Remote reading, solid state instruments.

Illinois
Airguide Instrument Co.
2210 Wabansia Avenue
Chicago, IL

Raeco, Inc.
550 Armory Drive
South Holland, IL

Iowa
J-Tec Associates, Inc.
317 Seventh Avenue, S.E.
Cedar Rapids, IA 52401

Maryland
Belfort Instrument Co.
1600 S. Clinton Street
Baltimore, MD 21224
Indicating and recording instruments.

Davis Instrument Mfg. Co., Inc.
521 E. 36th Street
Baltimore, MD 21218

Simerl, R.A., Instrument Div.
238 West Street
Annapolis, MD 21401
Wind measuring devices

Massachusetts

Alden Electronic & Impulse Recording Equipment Co.
Washington Street
Westboro, MA

Cape Cod Wind & Weather Indicators
625 Main Street
Harwich Port, MA

Datamarine International, Inc.
Commerce Park Road
Pocasset, MA

Downeaster Manufacturing Co.
574 Route 6-A
Dennis, MA
Wind and weather indicators.

General Eastern Instruments
50 Hunt Street
Watertown, MA 02172
Industrial hygrometers.

Maximum Inc.
42 South Avenue, Suite 34
Natick, MA 01760
Wind and weather instrumentation, aneroid barometers.

M.C. Stewart
Box 338
Ashburnham, MA
Wind, evaporation, forest fire, rain instruments.

Robert E. White Instruments
51 Commercial Wharf
Boston, MA

Wright & Wright, Inc.
P.O. Box 1728
Oak Bluffs, MA
Automatic visual range instruments.

Michigan

Heath Company
Benton Harbor, MI 49022
Electronic weather instrument kits.

R.M. Young Co.
2801 Aero-Park Drive
Traverse City, MI 49684
Sensitive wind measuring equipment.

Testrite, Inc.
887 DeGurse
Marine City, MI

Minnesota

Kavouras, Inc.
6301 34 Avenue, S.
Minneapolis, MN

Sparsa Products, Inc.
417 6th Avenue, N.E.
Osseo, MN
Thunderstorm trackers and monitors, weather alarms.

Missouri

Seiler Instrument & Mfg. Co.
172 East Kirkham Avenue
St. Louis, MO
Balloon theodolites and transits.

Nevada

Bahn-Fair, Inc.
P.O. Box 1735
Carson City, NV

New Hampshire
Hollis Observatory
1-A Pine Street
Nashua, NH 03060

Natural Power Inc.
New Boston, NH 03070

New Jersey
Edmund Scientific
Edscorp Building
Barrington, NJ 08007

Harris & Mallow Clocks, Inc.
651 New Hampshire Avenue
Lakewood, NJ

Science Associates, Inc.
230 East Nassau Street
Princeton, NJ 08540
*Temperature, humidity, pressure, wind,
sunshine and precipitation instruments.*

New Mexico
Thunder Scientific Corp.
623 Wyoming, S.E.
Albuquerque, NM
Relative humidity instruments.

New York
Beukers Laboratories, Inc.
Flowerfield Building
St. James, NY
Automatic data processing.

Cardion Electronics
Long Island Expressway
Woodbury, NY
*Wind, temperature, and humidity
measuring instruments.*

Climatronics Corp.
140 Wilbur Place
Bohemia, NY 11716
Instruments and sensors.

Conkling Laboratories
5432 Merrick Road
Massapequa, NY
Degree-day instruments.

Cuckoo Clock Manufacturing Co.
32-40 West 25th Street
New York, NY
Weather instruments.

Phys-Chemical Research Corp.
36 West 20th Street
New York, NY 10011

Servo Corp. of America
111 New South Road
Hicksville, NY
*Radiotheodolites, ranging systems,
tracking pedestals.*

Teledyne Gurley
P.O. Box 88
Troy, NY 12181

Watrous & Co., Inc.
172 Euston Road
Garden City, NY 11530

Weksler Instruments Corp.
80 Mill Road
Freeport, NY 11520
Thermometers, psychrometers.

Ohio
Electric Speed Indicator Co.
12234 Triskett Road
Cleveland, OH 44111
Wind speed and direction indicators.

Inservco, Inc.
114 Commerce Avenue
LaGrange, OH
Solar intensity meters.

Technology, Inc.
1115 Talbott Tower
Dayton, OH
Wind shear instruments.

Wong Laboratories
3357 Madison Road
Cincinnati, OH
Research and development.

Oregon
Hinds International, Inc.
P.O. Box 4327
Portland, OR 97208
Digital weather stations.

Leupoid & Stevens, Inc.
P.O. Box 688
Beaverton, OR
Weather, rain, and snow instruments.

Met One, Inc.
P.O. Box 1937
Grants Pass, OR

Pennsylvania
Abele Industrial
R.D. 2, Box 173
Zelienope, PA
Rain gauges, indicators, wind instruments.

Environmental Tectonics Corp.
12 County Line Industrial Park
Southampton, PA 18966
Psychrometers.

H-B Instrument Company
4303 North American Street
Philadelphia, PA 19140
Psychrometers, hygrometers, thermometers.

Kurt J. Lesker Co.
5641 Horning Road
Pittsburgh, PA

Princo Instruments
1020 Industrial Highway
Southampton, PA
*Thermometers, psychrometers,
mercurial barometers.*

Vista Scientific Corp.
85 Industrial Road
Ivyland, PA 18974
Temperature and humidity instruments.

Texas
Teledyne Geotech
3401 Shiloh Road
Garland, TX

Texas Electronics Inc.
Inwood Station, Box 7225
Dallas, TX
Electronic remote weather instrument sensors.

Weather Scan, Inc.
Throckmorton Highway
Olney, TX
Weather equipment for CATV industry.

Utah
Laird Telemedia, Inc.
2424 South 2570 West
Salt Lake City, UT

Virginia
Atlantic Research Corp.
5390 Cherokee Avenue
Alexandria, VA
Meteorological sounding rocket systems.

Teledyne Hastings-Raydist
P.O. Box 1275
Hampton, VA
Air velocity instruments.

Canada
Norstream Intertec, Inc.
P. O. Box 1507
St. Catherine's, Ontario

Glossary

absolute humidity—A ratio of the quantity of water vapor present per unit volume of air, usually expressed as grams per cubic meter or grains per cubic foot. This ratio is of limited value to the meteorologist because slight changes in atmospheric pressure or temperature alter the amount of air and vapor in a specific volume, thus changing the absolute humidity even though the amount of moisture in the air has not changed.

active front—A front that produces appreciable cloudiness and precipitation.

anemometer—An instrument for measuring the force or speed of the wind.

anvil cloud—The popular name of a heavy cumulus or cumulonimbus cloud having an anvil-like formation of cirrus clouds in its upper portions. If a thunderstorm is seen from the side, the anvil form of the cloud mass is usually noticeable.

arctic front—The zone of discontinuity between the extreme cold air of the Arctic regions and the cool polar air of the northern Temperate Zone.

blizzard—A violent, intensely cold wind laden with snow.

buildup—A cloud with considerable vertical development.

ceiling—The height above the Earth's surface of the lowest layer of clouds of obscuration phenomena that is reported as broken, overcast or obscured and not classified as thin or partial.

Celsius scale—A centigrade temperature scale originally based on 0° for the boiling point of water and 100° as the freezing point of water, (i.e.), an inverted centigrade scale. It is now used interchangeably with the centigrade scale, with 0° freezing and 100° boiling temperatures.

cloud bank—A mass of clouds, usually of considerable vertical extent, stretching across the sky on the horizon, but not extending overhead.

cloudburst—A sudden and extremely heavy downpour of rain; frequent in mountainous regions where moist air encounters orographic lifting.

cold wave—A rapid and marked fall of temperature during the cold season of the year. The National Weather Service applies this term to a fall of temperature in 24 hours equaling or exceeding a specified number of degrees and reaching a specified minimum temperature or lower. Specifications vary for different parts of the country and for different periods of the year.

conduction—Conduction is the transfer of heat by contact. Air is a poor conductor of heat, therefore molecular heat transfer (conduction) during the course of a day or night affects only two or three feet of air directly. Wind and turbulence, however, continuously bring fresh air into contact with the surface and distribute the warmed or cooled air throughout the atmosphere.

convection—Although frequently used in physics to denote a complete atmospheric current, in meteorology *convection* refers to vertical air motion. The horizontal air movement that completes an air current is called *advection*.

cooling processes—Air temperature is decreased by all or any of the following processes: (1) *nocturnal cooling:* The Earth continuously radiates its heat outward toward space. During the night the loss of radiant energy lowers the temperature of the Earth's surface. The air temperature is thereafter reduced by conduction.
(2) *advection cooling:* When the windflow is such that cold air moves into an area previously occupied by warmer air, the temperature of the air over the area is decreased. Also, warm air advection over a colder surface will result in conductive cooling of the lower air layers.
(3) *evaporative cooling.* When rain or drizzle falls from clouds, the evaporation of the water drops cools the air through which these drops are falling.
(4) *adiabatic cooling.* The process by which air cools due to a decrease in pressure. If air is forced upward in the atmosphere, the resulting decrease in atmospheric pressure surrounding the rising air allows the air to expand and cool adiabatically. Weather produced by lifting processes, such as frontal weather, convective and orographic thunderstorms, and upslope fog are the result of adiabatic cooling.

depression A cyclonic (low pressure) area.

diurnal—Actions completed within 24 hours, or pertaining to day time.

equinox—The moment, occurring twice each year, when the sun, in its apparent annual motion among the fixed stars, crosses the celestial equator; so called because then the night is *equal* to the day, each being 12 hours long over the whole Earth. The *autumnal equinox* occurs on or about September 22 when the sun is traveling southward; the *vernal equinox* on or about March 21, when the sun is moving northward.

gradient—The rate of increase or decrease in magnitude such as a pressure or temperature gradient. When a horizontal pressure gradient exists, the direct force exerted by the area of higher pressure is called the *pressure gradient force.* When used to describe a wind (*gradient wind*), gradient refers to winds above the influence of terrestrial friction—normally above 2000 or 3000 feet—where only pressure gradient force is affecting the speed of the wind.

greenhouse effect—This term is derived from the effect of the glass roof on a greenhouse, which transmits high-frequency insolation but blocks the passage of terrestrial radiation from within the glass enclosure. The greenhouse effect caused by clouds and impurities in the atmosphere is most noticeable at night when they reduce the nocturnal cooling of the Earth.

gust—Rapid fluctuations in wind speed with a variation of 10 knots or more between peaks and lulls.

horse latitudes—The subtropical high pressure region at approximately 30° latitude, characterized by calm, or light, variable winds.

hot wave—A period of abnormally high temperatures, usually lasting three or more consecutive days during each of which the maximum temperature is 90°F or over.

humidity—A general term to denote the water vapor content of the air.

inclination—The angle of the wind with respect to the isobar at the point of observation (usually between 20° and 30° at the surface).

lapse rate—A change in value expressed as a ratio, generally used with temperature changes vertically; (i.e.), 2°C per 1000 feet in the standard atmosphere.

mean sea level—In the United States, the average height of the surface of the sea for all stages of the tide during a 19-year period.

mesometeorology—The study of atmospheric phenomena such as tornadoes and thunderstorms that occur between meteorological stations or beyond the range of normal observation from a single point; (i.e.), on a scale larger than that of *micrometeorology,* but smaller than the cyclonic (*syntopic*) scale.

micrometeorology—The study of variations in meteorological conditions over very small areas, such as hillsides, forests, river basins, or individual cities. Also called *micro-climates* (Chapter 8).

natural air—Air as found in the atmosphere, containing water vapor and other impurities.

radiation—Electromagnetic waves traveling at 186,000 miles per second, many of which may be visible as light. Cosmic rays, gamma rays, X-rays, ultraviolet rays, visible light rays, infrared rays, and radio waves are some common types of radiation.

rate of evaporation—Rate of evaporation is the measured time for the changing state of a substance from a liquid to a gas. The quantity of vapor that will escape from a liquid surface into the air is primarily governed by the temperature of the liquid, the amount of vapor already in the air, and the speed of air movement over the liquid surface. Thus, much water will evaporate from the Great Lakes into the cold, dry winter air above it, even though the air cannot support the moisture in the vapor state. The resulting condensation forms a dense evaporation fog over the lakes.

secondary—A small area of low pressure on the border of a large (primary) area. The secondary may develop into a vigorous cyclone while the primary center disappears.

secondary circulation—In this wind classification category, many authorities include only migratory anticyclones (highs) and cyclones (lows). Such wind patterns as land and sea breezes, mountain and valley breezes, eddies, and Foehn winds are then classified as *local winds.*

selective absorption—Substances are selective in the radiation frequencies that they will absorb and radiate. The gases and impurities of which air is composed absorb and radiate only a few of their incident radiation frequencies. For example, water vapor absorbs several of the lower frequency radiation waves of terrestrial radiation, but is transparent to the high frequency radiation from the sun.

solstice—The time of year when the direct ray of the sun is farthest from the Equator. The approximate dates are: summer solstice, June 22; winter solstice, December 22.

spread—The difference between the temperature of the air and the dew point of the air, expressed in degrees. Although there is a definite relationship between spread and relative humidity, a spread of 5°F between 90°F and 85°F produces a significantly different relative humidity from the same spread between 65°F and 60°F.

squall—A line of thunderstorms, generally continuous, across the horizon. The squall line is associated with prefrontal activity.

subsidence inversion—An inversion layer that forms near a center of high pressure where the entire column of air is descending (subsiding) toward the surface. As the layers descend, they are compressed by the inflow of fresh air aloft. Compression heats the subsiding air layer and often the layer becomes warmer at the base levels than at the upper levels. The resulting increase in temperature through the layer is a *subsidence inversion.* Haze layers often develop below these inversion layers. Over large, industrialized areas they cause *smog.*

syntopic—That which presents a general view of the whole; as a syntopic weather map or a syntopic weather situation in which the major weather phenomena over a large geographical area are depicted or discussed.

temperature lag—Although the sun is directly overhead at noon, incoming radiation continues to exceed reradiation from the Earth until after 2 p.m. local standard time. Thus, the diurnal surface temperature increase reaches a maximum in the midafternoon. Also, seasonally, the sun is highest in the Northern Hemisphere at the summer solstice (June 22), but the long hours of daylight and relatively direct incident radiation cause the summer temperatures to continue increasing into July and August.

weather—The state of the six meteorological elements in the atmosphere: air temperature, humidity, clouds, precipitation, atmospheric pressure, and wind. The term *weather* also has several specialized meanings in meteorology. It may refer to only the forms of precipitation in the atmosphere at the time of a meteorological observation, both the precipitation and obstructions to vision (fog, haze, smoke, dust, etc.), or all forms of atmospheric phenomena.

wind velocity—The speed and direction of the wind.

Index

Index

Other Bestsellers From TAB

☐ **TIME GATE: HURTLING BACKWARD THROUGH HISTORY—Pellegrino**

Taking a new approach to time travel, this totally fascinating history of life on Earth transports you backward from today's modern world through the very beginnings of man's existence. Interwoven with stories and anecdotes, and illustrated with exceptional drawings and photographs, this is history as it should always have been written! It will have you spellbound from first page to last! 288 pp., 142 illus. 7" × 10".

Paper $16.95   **Book No. 1863**

☐ **VIOLENT WEATHER: HURRICANES, TORNADOES AND STORMS—Gibilisco**

What causes violent storms at sea? Hurricane force winds? Hail the size of grapefruit? Blinding snowstorms and tornadoes? The answers to all these and many more questions on the causes, effects, and ways to protect life and property from extremes in weather are here in this thoroughly fascinating study of how extremes in weather violence occur. 272 pp., 192 illus. 7" × 10".

Paper $13.95   **Book No. 1805**

☐ **BLACK HOLES, QUASARS AND OTHER MYSTERIES OF THE UNIVERSE**

Discover the awesome mysteries and complexities of our universe: black holes, quasars, pulsars, quarks, matter and antimatter, superluminal motions, creation theories, the search for intelligent life on other planets or in other galaxies, the possibilities of travel at speeds faster than light, and more. Includes 16 spectacular color photos of space phenomena! 208 pp., 125 illus. 8 color pages. 7" × 10".

Paper $13.50   **Book No. 1525**

☐ **333 SCIENCE TRICKS AND EXPERIMENTS—Brown**

Here is a delightful collection of experiments and "tricks" that demonstrate a variety of well-known, and not so well-known, scientific principles and illusions. Find tricks based on inertia, momentum, and sound projects based on biology, water surface tension, gravity and centrifugal force, heat, and light. Every experiment is easy to understand and construct . . . using ordinary household items. 208 pp., 189 illus.

Paper $9.95   **Book No. 1825**

☐ **$E = mc^2$: PICTURE BOOK OF RELATIVITY**

It sounds complicated, but with this exceptional and easy-to-follow handbook, *anyone* child or adult, can grasp the real meaning of Einstein's theories. It's an enlightening, delightfully illustrated look at relativity minus the difficult mathematical equations and confusing scientific jargon. You'll be able to clearly understand how he reached his conclusions and the experiments that proved his theories correct. 128 pp., 138 illus. 7" × 10".

Paper $9.95   **Book No. 1580**

☐ **ASTRONOMY AND TELESCOPES: A BEGINNER'S HANDBOOK**

If you're fascinated with space phenomena and the wonders of our universe, here's you chance to discover the solar system, the Milky Way and beyond . . . to follow the path of meteors, asteroids, and comets . . . *even* to build your own telescope to view the marvels of space! You'll get an amazing view of the brilliance of outer space in 16 pages of full-color photos! 192 pp., 194 illus. 7" × 10".

Paper $14.95   **Book No. 1419**

*Prices subject to change without notice.

Look for these and other TAB BOOKS at your local bookstore.

TAB BOOKS Inc.
P.O. Box 40
Blue Ridge Summit, PA 17214

Send for FREE TAB Catalog describing over 900 current titles in print.